Stochastic methods of
operations research

Stochastic methods of operations research

J. KOHLAS

Director of the Institute for Automation and Operations Research at the University of Fribourg

TRANSLATED BY A. SCHMIDT

CAMBRIDGE UNIVERSITY PRESS

CAMBRIDGE

LONDON   NEW YORK   NEW ROCHELLE
MELBOURNE   SYDNEY

Published by the Press Syndicate of the University of Cambridge,
The Pitt Building, Trumpington Street, Cambridge CB2 1RP
32 East 57th Street, New York, NY 10022, USA
296 Beaconsfield Parade, Middle Park, Melbourne 3206, Australia

Orginally published in German as *Stochastische Methoden des Operations
Research* by B. G. Teubner, Stuttgart, 1977 and © B. G. Teubner, Stuttgart
1977. First published in English with permission of B. G. Teubner Verlag,
Stuttgart, by Cambridge University Press 1982 as *Stochastic methods of
operations research.*

English Edition © Cambridge University Press 1982

Printed in Great Britain at the University Press, Cambridge

Library of Congress catalogue card number: 81-21574

British Library Cataloguing in Publication Data
Kohlas, J.
Stochastic methods of operations
research.
1. Operations research 2. Stochastic
processes.
I. Title   II. Stochastische Methoden
des Operations Research. *English*
001.4'24   T57.6
ISBN 0 521 23899 4 hard covers
ISBN 0 521 28292 6 paperback

Contents

v

Contents

Preface

Operations research deals with the mathematical analysis of techno-economical systems and problems. In this context one invariably has to deal with more or less pronounced uncertainty and indeterminacy. Very often one can ignore the uncertainty and work with estimates, means or expected values. There are, however, problems whose very nature is determined randomly. The essence of the problem would be lost if one tried to eliminate the random factor. In such cases the problem has to be tackled with probabilistic or stochastic methods.

Since these problems almost always involve dynamical or time evolution, they fall into the area of stochastic processes. The fundamental tools for the treatment of stochastic problems of operations research are renewal theory and Markov chains (Chapters 2 and 3). Important applications of these appear in queueing systems (Chapter 4) and in dynamical optimization (Chapter 5). The numerical treatment of stochastic problems of operations research is of the greatest practical importance and is based on simulation and Monte-Carlo methods (Chapter 6). The important extension of both renewal processes and Markov chains is furnished in the theory of semi-Markov models (Chapter 7). We do not attempt here to give a comprehensive and complete account of these individual fields. The principal aim is to provide an introduction to those ideas which are most fundamental and characteristic for each subject area. Naturally, the stochastic methods presented here are based on general probability theory. This is the reason for the concise introduction to probability theory at the beginning of Chapter 1 which contains its most essential results, most of them presented without proofs. Readers familiar with probability theory may skip this chapter.

The logical dependence of the individual chapters is shown in the following diagram:

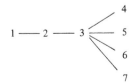

Preface

The use of stochastic methods can be found in almost all applications of operations research and in particular in problems of maintenance and service, of replacement, renewal, and reliability, in operating problems as well as stock-keeping, transport and traffic flow problems. The examples included in the text convey an idea of likely applications. For a more detailed presentation of applications we refer the reader to specialist textbooks covering these areas.

The ends of proofs are marked by ■, the ends of examples by □. Formulae are numbered consecutively in each section and are referred to simply as (k). A reference to formula (k) of a different Section j in Chapter i is $(i.j.k)$. Exactly the same system applies to theorems. Examples are marked (a), (b), (c), ... in every section and are referred to in exactly the same way as formulae or theorems. At the end of each chapter there are notes which contain references to relevant literature. The bibliography at the end of the book has been arranged by chapter. It only represents a small selection. We have concentrated mainly on textbooks and have included articles from journals only if the material in them has not yet found its way into the textbooks. Readers who would like to study one of the areas in greater detail are referred to the reference sections of the books listed in the bibliography.

I would like to thank my colleagues, Professor P. Kall who encouraged me to write this book and who assisted me with his valuable advice, Professor W. Runggaldier with whom I had my stimulating discussions, and W. Weidert, dipl.math. and Nguyen Dac Hoa, lic.oec., for their help with checking the manuscript, Mrs Rocher for her careful typing and the publishers B. G. Teubner for accepting this text for their LAMM-series.

Fribourg   J. Kohlas
Switzerland 1977

Preface to the English edition

It was with great pleasure that I received the news that Cambridge University Press were planning to publish an English translation of my book. At the beginning of the developments of operations research, German speaking students, including myself, greatly benefited from German translations of fundamental books on operations research, I hope that now in the reverse direction my book will be equally useful to English speaking readers.

I would like the reader to interpret the title of the book properly: it is about *methods* (more precisely stochastic methods) in operations research and not about operations research itself. Methods are tools to solve problems. Knowledge of tools is a necessary but not a sufficient condition for treating problems of operations research. Be careful to understand the physical aspects of the operations you want to study and research equally well as the methods you learn in this book.

The content of the original German edition has, in this English translation, been supplemented by a section on numerical computation of transient solutions to continuous parameter Markov chains (Section 3.5) and a chapter on semi-Markov models (Chapter 7). Otherwise nothing has been changed, except that we tried to eliminate the few errors found in the German edition, and we hope that we have not introduced too many new ones.

My thanks go in the first place to the translator, Mrs Annelise Schmidt. I hope she will accept the greatest compliment that I can give: that the translation reads as though it was written by myself, only in a better style than my own. This means of course that in the English edition as well as the German one I am to blame for all the methodological or pedagogical deficiencies that may be found. I would also like to thank my collaborators who checked the text, especially: Mr Magnus Branderberger, Mr Jacques Pasquier, Mr Uong Dinh Manh. Mrs Gabrielle Rocher who typed the German edition, was also kind enough to accept typing the supplementary parts of the English edition, and Mrs Ruth Leu also helped in preparing the English version. For this I want to thank them both. Finally I thank Cambridge University Press for preparing an English edition of my book.

Fribourg, Switzerland   J. Kohlas
Spring 1981

1
Probability theory

1.1. Probability spaces

The common theme in the following chapters is the description and investigation of random processes. The most important concepts in it are those of *probability* and of *random variable*. In this first chapter we introduce these fundamental concepts and collect the main results of probability theory. The presentation has been chosen in accordance with the requirements of the subsequent chapters. The first chapter will therefore list the fundamental results of probability theory partly without giving proofs.

Any particular test, experiment, any particular operation or process leads *a priori* to a set of *possible outcomes*. Abstractly this is described by a set Ω whose elements ω represent the possible outcomes of the random experiment under consideration. The possible outcomes are called *sample values* or *elementary events* and the set Ω is the *sample space*.

Examples

(a) *Coin tossing.* If a coin is tossed *once* there are two possible sample values or elementary events, i.e. head (H) or tail (T). Accordingly the sample space $\Omega^{(1)}$ consists of the *two* elements H and T; $\Omega^{(1)} = \{H, T\}$. If the experiment to be considered consists of tossing a coin twice then the combination of the events of the two tosses yields *four* possible sample values, i.e. $\Omega^{(2)} = \{HH, HT, TH, TT\}$. With n tosses the sample space $\Omega^{(n)}$ consists of all possible n-tuples of the symbols H and T. It contains therefore 2^n elements. □

(b) *Throwing dice.* When throwing a die the possible sample values correspond to the number of dots on the upper face of the die. With an ordinary die we therefore obtain $\Omega = \{1, 2, 3, 4, 5, 6\}$. □

(c) *Life expectancy.* The possible life span of a strip light, for example, can be represented by non-negative numbers x. The sample values are thus the non-negative real numbers: $\Omega = [0, \infty)$. It does not matter if the sample space is extended by including additional and, in fact, non-realizable sample values. In this example one might therefore equally well choose the set of all real

1

numbers $\mathbb{R} = (-\infty, \infty)$ as sample values. If one looks at the life of n strip lights then the n-tuples $(x_1, x_2, \ldots, x_n)$ are to be considered as sample values, where x_i is the life span of the ith strip light. In this case the sample space is the n-dimensional space $\mathbb{R}^n$ of all n-tuples of real numbers. $\mathbb{R}$ and $\mathbb{R}^n$ are frequently-used and important sample spaces.   □

When examining random experiments one is often less interested in single sample values than in particular *events*. A particular event occurs precisely when during the experiment a sample lies in a particular and specified set of sample values. An event A is thus nothing but a *subset of the sample space*. If in the course of an experiment the sample value $\omega \in \Omega$ occurs then the event A has occurred if $\omega \in A$, and the event A has not occurred if $\omega \notin A$. The sample space Ω, considered as a subset of itself, is a particular event: the *certain event* which always occurs. Furthermore, the *empty set* ϕ, the subset which contains no elements, is always considered to be an event: the *imposs-ible event*.

Further events can be constructed from given events. If A is an event then the *opposite of A*, A^c, can be constructed, i.e. the event which occurs exactly when A does *not* occur. A^c is the set theoretic complement of A, i.e. A^c contains precisely those sample values which are not contained in A,

$$A^c = \{\omega : \omega \notin A\}. \tag{1}$$

Note that $(A^c)^c = A$, $\Omega^c = \phi$ and $\phi^c = \Omega$.

If A and B are two events, an event C can be constructed which occurs exactly if both A and B happen. C is the event 'A and B'. Set theoretically C is the *intersection $A \cap B$*, i.e. the set of all sample values which belong to A and to B at the same time:

$$C = A \cap B = \{\omega : \omega \in A, \omega \in B\}. \tag{2}$$

Two events A and B are called *incompatible* or *disjoint* if $A \cap B = \phi$, i.e. if they cannot occur simultaneously. If $A \subset B$, i.e. if A is a subset of B then B certainly occurs if A occurs. It follows that $C = A \cap B = A$! If $I = \{1, 2, \ldots\}$ is a finite or countably infinite index set and if A_i is an event for all $i \in I$ then one can construct the event C equal to 'A_1 and A_2 and A_3 and $\ldots$':

$$C = \bigcap_{i \in I} A_i = \{\omega : \omega \in A_i \quad \text{for all} \quad i \in I\}. \tag{3}$$

If $A_1 \supset A_2 \supset A_3 \supset \ldots$, then (3) is also called $\lim_{i \to \infty} A_i$.

Furthermore, one can construct an event C out of two events A and B which occurs when *at least one* of the two events A and B occurs. Then C is the event 'A or B'. Set theoretically C is the *union $A \cup B$*, i.e. the set of the sample values which are contained in at least one of the two events A and B

$$C = A \cup B = \{\omega : \omega \in A \text{ and/or } \omega \in B\}. \tag{4}$$

If $A \subset B$ then it follows that $C = A \cup B = B$. Out of the $A_i, i \in I = \{1, 2, \ldots\}$ one can accordingly construct the event C equal to 'A_1 or A_2 or A_3 or ...':

$$C = \bigcup_{i \in I} A_i = \{\omega : \omega \in A_i \text{ for at least one } i \in I\}. \tag{5}$$

If $A_1 \subset A_2 \subset A_3 \subset \ldots$, then (5) is also denoted by $\lim_{i \to \infty} A_i$.

In general not all subsets of the sample space Ω are taken into consideration as events, but only a certain class $\mathscr{A}$ of subsets of Ω. $\mathscr{A}$ should, however, be large enough so that the above operations, carried out with events in $\mathscr{A}$, again result in events in $\mathscr{A}$. Therefore $\mathscr{A}$ has to satisfy the following conditions.

1. $\Omega \in \mathscr{A}, \phi \in \mathscr{A}$.
2. If $A \in \mathscr{A}$, then $A^c \in \mathscr{A}$,
3. If $A_i \in \mathscr{A}, i = 1, 2, \ldots$, is a finite or countably infinite sequence of events then $\cap_i A_i \in \mathscr{A}$ and $\cup_i A_i \in \mathscr{A}$.

It should be observed that this is not a minimal list of conditions; some of the conditions are consequences of the others. A class $\mathscr{A}$ which satisfies conditions 1 to 3 is called a σ-algebra (sigma algebra). The simplest but usually useless σ-algebra consists only of Ω and ϕ.

Examples

(d) *Coin tossing* (see Example (a)). In $\Omega^{(1)}$ (single toss) one is able to define at most four events, i.e. ϕ, {H}, {T} and $\Omega = \{H, T\}$; it is important to distinguish carefully between the element (sample) ω and the set (event) $\{\omega\}$, which consists of the single element ω. The four events form a σ-algebra. In $\Omega^{(2)}$ (tossing twice) there are already sixteen possible events which again form a σ-algebra. Examples are the events: 'first toss is a head' = {HH, HT}, 'second toss is a tail' = {HT, TT}, 'different results in both tosses' = {HT, TH}. □

(e) *Throwing dice* (see Example (b)). Again all subsets of $\Omega = \{1, 2, 3, 4, 5, 6\}$ form a σ-algebra. One may only be interested in the events 'even throw' = {2, 4, 6} and 'odd throw' = {1, 3, 5} which together with ϕ and Ω again form a σ-algebra. □

(f) *Countable sample spaces*. For sample spaces Ω with a finite or countably infinite number of sample values one often considers the σ-algebra consisting of *all* subsets of Ω. □

(g) *Borel algebras*. In the important sample space $\mathbb{R} = (-\infty, \infty)$ one would particularly like to look at events which are defined by open intervals (a, b), closed intervals $[a, b]$ and half-open intervals $(a, b]$ and $[b, a)$. There surely must exist a σ-algebra which contains these events, the one which consists of all subsets of $\mathbb{R}$. We now consider *all* σ-algebras which contain the above mentioned intervals and denote by $\mathscr{B}$ the class of subsets which are contained in all these σ-algebras. Obviously $\mathscr{B}$ satisfies the above conditions 1 to 3 and is therefore itself a σ-algebra which contains these intervals. Indeed, $\mathscr{B}$ is obviously the *smallest* σ-algebra which contains all intervals. $\mathscr{B}$ is called the *Borel algebra*; it contains all sets which can be obtained by forming countable intersections, unions and complements out of a countable number of intervals.

In the more general case of $\mathbb{R}^n$ the Borel algebra is defined in an analogous way as the smallest σ-algebra containing all n-dimensional 'intervals' of the form $(a_1 < x_1 < b_1; \ldots; a_n < x_n < b_n)$, as well as all half-open and closed versions of these $\qquad\qquad\qquad\qquad\qquad\qquad\qquad$ $\square$

In the next step we assign to the events of our σ-algebra $\mathscr{A}$ *probabilities* $P(A)$ for every $A \in \mathscr{A}$. The probabilities $P(A)$ are real numbers in the interval $[0, 1]$ which may not, however, be assigned arbitrarily to the events. The following conditions have to be satisfied.

1. $P(\Omega) = 1, P(\phi) = 0, 0 \leqslant P(A) \leqslant 1$ for all $A \in \mathscr{A}$.
2. For every $A \in \mathscr{A}, P(A^c) = 1 - P(A)$.
3. If $A_i \in \mathscr{A}, i = 1, 2, \ldots$, is a finite or countably infinite sequence of *disjoint* events $(A_i \cap A_j = \phi$, if $i \neq j)$, then

$$P\left(\bigcup_i A_i\right) = \sum_i P(A_i). \tag{6}$$

4. If for $A_i \in \mathscr{A}, i = 1, 2, \ldots$, either $A_1 \supset A_2 \supset A_3 \supset \ldots$ or $A_1 \subset A_2 \subset A_3 \subset \ldots$, then

$$\lim_{i \to \infty} P(A_i) = P(\lim_{i \to \infty} A_i). \tag{7}$$

Again some of these conditions could be obtained as a consequence of the others so that this list of conditions is also not minimal. A set function P which is defined on a σ-algebra $\mathscr{A}$ of a sample space Ω and which satisfies conditions 1 to 4 is called a *probability measure* and the triple $(\Omega, \mathscr{A}, P)$ a *probability space*.

Conditions 1 to 4 have the following significance: according to condition 1 the certain event is always to have probability 1 and the impossible event to have probability 0. All other events ought to possess probabilities between 0 and 1. According to condition 2 the probabilities of an event and its opposite should add up to 1 since it is certain that one of the two events will occur. According to condition 3 the probability that one of several incompatible events occurs should be equal to the sum of the probabilities of the individual events. Condition 3 is also called the *additivity condition*. Condition 4 ensures certain continuity properties of probabilities.

Conditions 1 to 4 have some important consequences. If $A \subset B$ then B is certain to occur whenever A occurs. Since $B = A \cup (A^c \cap B)$ and since A and $(A^c \cap B)$ are disjoint, it follows that $P(B) = P(A) + P(A^c \cap B)$. Since $P(A^c \cap B) \geqslant 0$, we have

$$P(A) \leqslant P(B), \quad \text{if } A \subset B. \tag{8}$$

If $C_i, i = 1, 2, \ldots$, is a family of *disjoint* events such that $\cup_i C_i = \Omega$, then $A = \cup_i (A \cap C_i)$ for every aribitrary event A and the events $A \cap C_i$ are again disjoint. From this we conclude

$$P(A) = \sum_i P(A \cap C_i). \tag{9}$$

In the following examples it is shown that it suffices to define the probabilities initially only for a particular subclass of events. As a consequence of conditions 1 to 4 the probabilities will then be uniquely determined for all other events in the σ-algebra. Probability theory, however, gives no information about *how* these initial probabilities are to be chosen: probability theory only starts *after* the choice of the probabilities has been made. The question of whether in a concrete application the correct probabilities have been chosen cannot be answered by probability theory but, at best, by mathematical statistics.

Examples

(*h*) *Coin tossing* (see Examples (*a*) and (*d*)). In general it is assumed that both head and tail are *equally likely* so that $P(\{H\}) = P(\{T\})$ (when tossing once). Since, according to condition 1, $P(\Omega) = P(\{H, T\}) = 1$ we conclude from condition 3 and from $\Omega = \{H\} \cup \{T\}, \{H\} \cap \{T\} = \phi, P(\Omega) = P(\{H\}) + P(\{T\}) = 1$ that $P(\{H\}) = P(\{T\}) = \frac{1}{2}$. $\qquad\square$

(*i*) *Countable sample spaces.* If the sample space Ω consists of a finite or countably infinite number of elements $\omega_i, i = 1, 2, \ldots$, and if one considers the σ-algebra which consists of all subsets of Ω, then it suffices to specify the probabilities $P(\{\omega_i\}) = p_i$ for $i = 1, 2, \ldots$ Of course, the p_i will have to satisfy $0 \leqslant p_i \leqslant 1$ and, since $\Omega = \cup_i \{\omega_i\}$,

$$\sum_i p_i = 1.$$

If A is an arbitrary event then condition 3 implies

$$P(A) = \sum_{\omega_i \in A} p_i.$$

The set function thus defined is obviously a probability measure. $\qquad\square$

(*j*) *Distribution functions.* In the sample space $\mathbb{R}$ with the Borel algebra (see Example (*g*)) one can define, to begin with, probabilities for the events $(-\infty, x] = \{\omega \in \mathbb{R}: \omega \leqslant x\}, -\infty < x < +\infty$. The probabilities of these events can be understood as a function of x, $P(\{\omega \leqslant x\}) = F(x)$. If $x_1 < x_2$, then $(-\infty, x_1] \subset (-\infty, x_2]$ so that, according to (8), $F(x_1) \leqslant F(x_2)$. Hence

$F(x)$ is a *monotonically non-decreasing* function. If furthermore $x_1 < x_2 < x_3 < \ldots$ and $\lim x_i = \infty$, then $(-\infty, x_1] \subset (-\infty, x_2] \subset (-\infty, x_3] \subset \ldots$ and $\lim_{i \to \infty} (-\infty, x_i] = (-\infty, \infty) = \mathbb{R}$. From conditions 1 and 4 we obtain $\lim_{x \to \infty} F(x) = 1$. Similarly $x_1 > x_2 > x_3 > \ldots$ and $\lim_{i \to \infty} x_i = -\infty$ imply $(-\infty, x_1] \supset (-\infty, x_2] \supset (-\infty, x_3] \supset \ldots$ and $\lim_{i \to \infty} (-\infty, x_i] = \phi$. It follows that $\lim_{x \to -\infty} F(x) = 0$. If finally the x_i converge to x *from above* $(x_i \downarrow x)$ then it follows that $\lim_{i \to \infty} (-\infty, x_i] = (-\infty, x]$ and thus $\lim_{x' \downarrow x} F(x') = F(x + 0) = F(x)$. Thus $F(x)$ is *continuous from the right*.

A function $F(x)$, defined for $-\infty < x < \infty$, which satisfies the conditions

$$F(x_1) \leqslant F(x_2) \qquad \text{for } x_1 \leqslant x_2,$$

$$\lim_{x \to \infty} F(x) = 1, \quad \lim_{x \to -\infty} F(x) = 0,$$

$$\lim_{x' \downarrow x} F(x') = F(x + 0) = F(x),$$

is called a (one-dimensional) *distribution function*. It can be shown that every distribution function determines a *unique* probability measure on the Borel algebra; i.e. from one distribution function we can determine the probability of every event in the Borel algebra. It is therefore sufficient to specify a distribution function for the sample space $\mathbb{R}$ in order to determine the probability structure.

We do not want to prove this in full generality but shall only show how the probabilities can be computed for the intervals, (a, b), $[a, b]$, $(a, b]$ and $[a, b)$. Let $x_i, i = 1, 2, \ldots$, be a monotonically increasing sequence which converges to a. Then $\lim_{i \to \infty} (x_i, b] = [a, b]$ and therefore (by condition 4) $P([a, b]) = \lim_{i \to \infty} P((x_i, b])$. If furthermore x_i is a monotonically increasing sequence converging to b then $\lim_{i \to \infty} (a, x_i] = (a, b)$, $\lim_{i \to \infty} [a, x_i] = [a, b)$ and therefore $P((a, b)) = \lim_{i \to \infty} P((a, x_i])$ and $P([a, b)) = \lim_{i \to \infty} P([a, x_i])$. We denote the limit of the function F at a point x when approached *from below* by $F(x - 0) = \lim_{x' \uparrow x} F(x')$. All that is left is the computation of the probability for intervals of the form $(a, b]$. Since $(-\infty, b] = (-\infty, a] \cup (a, b]$, $F(b) = F(a) + P((a, b])$ (condition 3). It follows that $P((a, b]) = F(b) - F(a)$. Summarizing, we obtain the following identities:

$$P((a, b]) = F(b) - F(a),$$

$$P([a, b]) = F(b) - F(a - 0),$$

$$P((a, b)) = F(b - 0) - F(a),$$

$$P([a, b)) = F(b - 0) - F(a - 0).$$

In general $F(x - 0)$ is quite different from $F(x)$. Distribution functions are not usually continuous from the left unless they are continuous. In this case, of course, $F(x - 0) = F(x)$ for all x and the four probabilities above all have the same value $F(b) - F(a)$.   □

If A and B are two events then we may notice during the course of a test that the event B has occurred. This will generally mean that the probability of the occurrence of A will change. This new probability is called the *conditional probability* $P(A|B)$ of A *given* B. This conditional probability is defined by

$$P(A|B) = \frac{P(A \cap B)}{P(B)}, \qquad (10)$$

provided that $P(B) > 0$. The conditional probability $P(A|B)$ is not defined if $P(B) = 0$. This definition has the following intuitive explanation: if one knows that B has occurred then the sample ω satisfies $\omega \in B$. Event A can only have occurred if $\omega \subset A \cap B$. The probability that A has occurred can therefore be measured by the relative probability of $A \cap B$ with respect to B. The following properties of conditional probabilities can easily be verified:

$$P(\Omega|B) = P(B|B) = 1, \quad P(\phi|B) = 0; \qquad (11)$$

$$P(A^c|B) = 1 - P(A|B). \qquad (12)$$

If furthermore the $A_i \in \mathscr{A}, i = 1, 2, \ldots,$ are *disjoint* then

$$P\left(\bigcup_i A_i | B\right) = \sum_i P(A_i|B). \qquad (13)$$

If finally $A_1 \supset A_2 \supset A_3 \supset \ldots$ or $A_1 \subset A_2 \subset A_3 \subset \ldots,$ then $(A_1 \cap B) \supset (A_2 \cap B) \supset \ldots$ or $(A_1 \cap B) \subset (A_2 \cap B) \subset \ldots$ respectively and therefore

$$\lim_{i \to \infty} P(A_i|B) = P(\lim_{i \to \infty} A_i|B). \qquad (14)$$

Combining all this we observe that the set function $P(A|B)$ is, for every fixed event B with $P(B) > 0$, and as a function of $A \in \mathscr{A}$, a probability measure on $\mathscr{A}$ which satisfies the above conditions 1 to 4.

This implies in particular that relations analogous to (8) and (9) also hold for conditional probabilities. Relation (9) corresponds to

$$P(A|B) = \sum_i P(A \cap C_i|B), \qquad (15)$$

where the $C_i, i = 1, 2, \ldots,$ are all disjoint and satisfy $\bigcup_i C_i = \Omega$.

If $P(A|B) = P(A)$, then it follows from (10) that

$$P(A \cap B) = P(A)P(B) \qquad (16)$$

and from this, if $P(A) > 0$, it also follows that $P(B|A) = P(B)$. Two events A and B, which satisfy (16), are called (mutually) *independent*. For independent events the probability of one event does not change if the other has occurred.

Example

(k) *Coin tossing* (see Example (h)). Suppose that, tossing the coin twice, we have obtained H at the first toss, i.e. the event $B = \{HH, HT\}$ has occurred. Given this event, we would like to know the conditional probability for obtaining T at the second toss, i.e. for $A = \{HT, TT\}$. Since $A \cap B = \{HT\}$, $P(A \cap B) = \frac{1}{4}$, $P(B) = \frac{1}{2}$ we conclude that $P(A|B) = \frac{1}{4}/\frac{1}{2} = \frac{1}{2}$. But this is also the probability of A and therefore the two events A and B are *independent*. In this way it can be shown that all possible outcomes of the two tosses are mutually independent.   □

We finish with a few useful formulae for conditional probabilities. Let $A_i \in \mathscr{A}, i = 1, 2, \ldots$, be a finite or countably infinite sequence of *disjoint* events so that $\cup_i A_i = \Omega$ and $P(A_i) > 0$ for all i, and let $A \in \mathscr{A}$ be an arbitrary event. Then $A = \cup_i (A \cap A_i)$ and the events $A \cap A_i$ are disjoint. From (10) and condition 3 for probability measures, it therefore follows that

$$P(A) = \sum_i P(A_i)P(A|A_i). \tag{17}$$

Equation (17) is called the formula of *total probability*.

Let A, B, C be three events with $P(A) > 0$, and $P(A \cap B) > 0$. Applying (10) twice we get

$$P(A \cap B \cap C) = P(A \cap B)P(C|A \cap B) = P(A)P(B|A)P(C|A \cap B). \tag{18}$$

If one divides both sides of (18) by $P(A)$, the outcome is the relation

$$P(B \cap C|A) = P(B|A)P(C|A \cap B). \tag{19}$$

More generally, if $A_i \in \mathscr{A}$ are events with $P(A_1) > 0, P(A_1 \cap A_2) > 0$, $P(A_1 \cap A_2 \cap A_3) > 0, \ldots$ (18), together with induction on n, yields

$$P(A_1 \cap A_2 \cap A_3 \cap \ldots \cap A_n) = P(A_1)P(A_2|A_1)P(A_3|A_1 \cap A_2)$$
$$\ldots P(A_n|A_1 \cap A_2 \cap \ldots \cap A_{n-1}). \tag{20}$$

The events $A_1, A_2, A_3, \ldots, A_n$ are called *mutually independent* (or simply independent), if, for all subsets of indices $i_1, i_2, \ldots, i_n$,

$$P(A_{i_1} \cap A_{i_2} \cap \ldots \cap A_{i_n}) = P(A_{i_1})P(A_{i_2}) \ldots P(A_{i_n}). \tag{21}$$

In this case we also have $P(A_i|A_j) = P(A_i)$ for all $i \neq j$ and the A_i, A_j are pairwise independent. Furthermore $P(A_i|A_{j_1} \cap \ldots \cap A_{j_m}) = P(A_i)$ for all $j_1, \ldots, j_m \neq i, m \leq n$.

1.2. Random variables

In the application of probability theory one works more often with random variables than with the underlying probability spaces. Intuitively a random variable is a quantity which adopts randomly a value from a particular range. Mathematically, random variables are defined as follows: if $(\Omega, \mathcal{A}, P)$ is a *probability space* and $X(\omega)$ a *real function* which is defined for every $\omega \in \Omega$ in such a way that the events $\{\omega : X(\omega) \leqslant x\}$ belong to $\mathcal{A}$ for every $x, -\infty < x < \infty$, then this function X is called a *random variable* (or, in the language of measure theory, an $\mathcal{A}$-measurable function).

If X is a random variable and $\mathcal{B}$ the Borel algebra in $\mathbb{R}$ (see Example (1.1.g)), then we can show that the event $\{\omega : X(\omega) \in B\}$ belongs to $\mathcal{A}$ for every $B \in \mathcal{B}$. These events are denoted by $X^{-1}(B)$. If one puts $P'(B) = P(X^{-1}(B))$ for every $B \in \mathcal{B}$ then P' becomes a *probability measure* on $\mathcal{B}$ and therefore $(\mathbb{R}, \mathcal{B}, P')$ a *probability space*. When considering a random variable X one may therefore, if one likes to, always restrict oneself to the probability space $(\mathbb{R}, \mathcal{B}, P')$ and forget about the original space $(\Omega, \mathcal{A}, P)$. Every random variable X on $(\Omega, \mathcal{A}, P)$ can also be interpreted as a random variable on $(\mathbb{R}, \mathcal{B}, P')$ by defining $X(x) = x$ for every $x \in \mathbb{R}$.

To simplify things we introduce the following notation which is used from now on: for the event $\{\omega : X(\omega) \in B\}$ we simply write $[X \in B]$ and for the probability of this event $P[X \in B]$. In the course of all further explanation we shall almost exclusively consider events which can be defined by conditions on random variables. Thus the underlying probability space $(\Omega, \mathcal{A}, P)$ is kept more and more in the background. From Chapter 2 onwards the underlying probability spaces are not explicitly mentioned any more, but implicitly assumed; strictly speaking, however, their existence has to be proved. We shall ignore this problem since the interest of applied probability theory lies elsewhere.

If one considers the events $[X \leqslant x]$ in the probability space $(\mathbb{R}, \mathcal{B}, P)$ for all real numbers x one obtains a function

$$F(x) = P[X \leqslant x], \qquad -\infty < x < \infty. \tag{1}$$

Example $(1.1.j)$ shows that $F(x)$ has to be a distribution function on $\mathbb{R}$. $F(x)$ is called the *distribution function of random variable X*. Conversely, every distribution function $F(x)$ determines a probability measure P on $\mathcal{B}$ and $(\mathbb{R}, \mathcal{B}, P)$ becomes a probability space according to Example $(1.1.j)$. Thus $X(\omega) = \omega$ for every $\omega \in \mathbb{R}$ is a *random variable* whose distribution function is just $F(x)$. Every distribution function therefore determines a corresponding random variable.

If $F(x) = 0$ for $x \leqslant 0$, i.e. $P[X \leqslant 0] = 0$ then the random variable X determined by the distribution function takes on negative values only with probability zero, i.e. the random variable X can in this case only assume *positive* values. Instead of the distribution function $F(x) = P[X \leqslant x]$ one can also consider the function $P[X > x] = 1 - F(x)$. In certain cases this can be quite useful.

If $g(x)$ is a real valued function defined for $-\infty < x < +\infty$ in such a way that, for every y, the set of x for which $g(x) \leqslant y$, lies in the Borel algebra, then $g(x)$ is a random variable by definition. In this case one speaks of a $\mathscr{B}$-measurable function. If furthermore X is a random variable on a probability space $(\Omega, \mathscr{A}, P)$ then $Y(\omega) = g(X(\omega))$ is again a random variable on the same space which we simply call $Y = g(X)$.

Most of the random variables introduced in the following chapters are either of *discrete* or *continuous* type. A random variable X of *discrete* type can only take on one of *finitely* or *countably infinitely* many values x_1, $x_2, \ldots$. Here we assume that these values have no finite accumulation point. The probabilities of the individual values are given by $P[X = x_i] = p_i$, $0 < p_i < 1, i = 1, 2, \ldots$. Since the event $[X \leqslant x]$ is the union of all events $[X = x_i]$ with $x_i \leqslant x$, the distribution function $F(x)$ of a random variable of discrete type satisfies

$$F(x) = P[X \leqslant x] = \sum_{x_i \leqslant x} p_i. \tag{2}$$

$F(x)$ is a *step function*; it only changes its value at countably many points $x_1, x_2, \ldots$ and has at the point x_i a *jump* of height $p_i, i = 1, 2, \ldots$,

$$F(x_i) - F(x_i - 0) = p_i = P[X = x_i]. \tag{3}$$

For $x \to \infty$ we have furthermore

$$\lim_{i \to \infty} F(x) = \sum_i p_i = 1. \tag{4}$$

The distribution function of a discrete random variable X is thus uniquely determined by the probabilities $\{p_i, i = 1, 2, \ldots\}$ of the possible values $x_1, x_2, \ldots$. The $p_i, i = 1, 2, \ldots$, together with the associated values x_i, $i = 1, 2, \ldots$, determine the *distribution* of the discrete random variable X. With discrete variables it is often more convenient to work with the probabilities p_i instead of the distribution function $F(x)$.

The set of possible values $\{x_1, x_2, \ldots\}$ which a discrete random variable can assume, is called the *range space* of the random variables X. The discrete random variables whose range space is equal to $\{0, 1, 2, 3, \ldots\}$ form an important special case: in this case $[X \leqslant n] = [X \leqslant n - 1] \cup [X = n]$ and $[X \geqslant n] = [X \geqslant n + 1] \cup [X = n]$, respectively. It thus follows that

$$p_n = P[X = n] = P[X \leqslant n] - P[X \leqslant n - 1]$$

$$= P[X \geqslant n] - P[X \geqslant n + 1]. \tag{5}$$

Examples

(a) *Degenerate random variables.* A random variable X which can only take on *one* value c, i.e. $X(\omega) = c$ for every $\omega \in \Omega$, is called *degenerate*. A degenerate random variable is equal to a deterministic quantity. We have $P[X = c] = 1$ and the distribution function is defined by $F(x) = 0$ for $x < c$ and $F(x) = 1$ for $x \geq c$. □

(b) *Bernoulli variables.* A random variable which can only take on the values 0 and 1 is called a *Bernoulli variable*. Its distribution is determined by $P[X = 1] = p, 0 < p < 1$, since, according to equation (4), $P[X = 0] = 1 - p$. □

(c) *Geometric distribution.* For $i = 0, 1, 2, \ldots$ we can define a distribution by $P[X = i] = p_i = (1 - p)p^i, 0 < p < 1, i = 0, 1, 2, \ldots$, since it can easily be checked that the sum of the p_i is equal to 1. Since the p_i form a geometric sequence this distribution is called *geometric*. □

(d) *Poisson distribution.* A discrete distribution can be determined by

$$P[X = i] = p_i = \frac{\lambda^i e^{-\lambda}}{i!}, \quad \lambda > 0, i = 0, 1, 2, \ldots,$$

since (4) can easily be verified by observing that the sum of $\lambda^i/i!$ is equal to e^λ. These distributions are called *Poisson distributions*. □

A random variable X is of *continuous* type if there exists a function $f(y) \geq 0$ for $-\infty < y < \infty$ such that the distribution function $F(x)$ can be represented in the form

$$F(x) = \int_{-\infty}^{x} f(y) \mathrm{d}y. \tag{6}$$

The function $f(y)$ can for all practical purposes be assumed to be *piecewise continuous* so that the integral in (6) can be understood as a Riemann integral. A distribution function $F(x)$ which is defined by (6) must necessarily be *continuous*, whereas not every continuous distribution function $F(x)$ can be represented in the form (6). Distribution functions of the form (6) are therefore called *absolutely continuous*. Since continuous distribution functions which are not absolutely continuous rarely occur in applications we shall drop the word 'absolute' for reasons of simplicity.

The function $f(x)$ is the *density function* of the random variable X with distribution function $F(x)$ given by (6). According to (6) the distribution function is uniquely determined by its density function. Furthermore we have, for $x_1 < x_2$ (see Example (1.1.j)),

$$P[x_1 < x_2] = F(x_2) - F(x_1) = \int_{x_1}^{x_2} f(y) \mathrm{d}y. \tag{7}$$

The same relation holds for $P[x_1 < X \leqslant x_2]$, $P[x_1 \leqslant X < x_2]$ and $P[x_1 \leqslant X \leqslant x_2]$ (see Example (1.1.*j*)). Equation (6) implies furthermore that

$$\int_{-\infty}^{\infty} f(y)dy = 1. \tag{8}$$

Examples

(e) *Uniform distributions.* These are defined for $a < b$ by the density function

$$f(y) = \begin{cases} 0 & \text{for } y < a, \, y > b, \\ 1/(b-a) & \text{for } a \leqslant y \leqslant b. \end{cases}$$

It follows from (6) that the distribution function is given by

$$F(x) = \begin{cases} 0 & \text{for } x < a, \\ (x-a)/(b-a) & \text{for } a \leqslant x \leqslant b, \\ 1 & \text{for } x > b. \end{cases}$$

This also shows that $f(y)$ satisfies condition (8). One speaks of a uniform distribution in the interval $[a, b]$. □

(f) *Exponential distributions.* These are defined for $\lambda > 0$ by the density function

$$f(y) = \begin{cases} 0 & \text{for } y < 0, \\ \lambda e^{-\lambda y} & \text{for } y \geqslant 0. \end{cases}$$

Equation (8) is satisfied and the corresponding distribution function is

$$F(x) = \begin{cases} 0 & \text{for } x < 0, \\ 1 - e^{-\lambda x} & \text{for } x \geqslant 0. \end{cases}$$ □

(g) *Normal distributions.* The density functions of normal distributions are

$$f(y) = \frac{1}{\sqrt{(2\pi)}\,\sigma} e^{-(y-\mu)^2/2\sigma^2}$$

with $-\infty < \mu < \infty$ and $0 < \sigma$. Let us start by verifying (8). If one puts $z = (y-\mu)/\sigma$ then

$$\frac{1}{\sqrt{(2\pi)}\,\sigma} \int_{-\infty}^{\infty} e^{-(y-\mu)^2/2\sigma^2} \, dy = \frac{1}{\sqrt{(2\pi)}} \int_{-\infty}^{\infty} e^{-z^2/2} \, dz.$$

Let I be the integral on the right hand side without the constant $1/\sqrt{(2\pi)}$. It remains to be shown that $I = \sqrt{(2\pi)}$. Consider

$$I^2 = \int_{-\infty}^{\infty}\int_{-\infty}^{\infty} e^{-(z_1^2+z_2^2)/2} \, dz_1 dz_2.$$

If one introduces the polar coordinates r, θ, defined by $z_1 = r \cos \theta, z_2 = r \sin \theta$, then $dz_1 dz_2 = r \, dr d\theta$ and $r^2 = z_1^2 + z_2^2$, hence

$$I^2 = \int_0^{2\pi} \int_0^\infty re^{-r^2/2} \, dr d\theta = 2\pi$$

and therefore $I = \sqrt{(2\pi)}$.

The distribution functions

$$F(x) = \frac{1}{\sqrt{(2\pi)}\,\sigma} \int_{-\infty}^x e^{-(y-\mu)^2/2\sigma^2} \, dy$$

cannot be integrated explicitly. On the other hand, using again the transformation $z = (y - \mu)/\sigma$, we obtain

$$F(x) = \frac{1}{\sqrt{(2\pi)}} \int_{-\infty}^{(x-\mu)/\sigma} e^{-z^2/2} \, dz = \Phi((x - \mu)/\sigma).$$

$\Phi(x)$ is the distribution function of the normal distribution with $\mu = 0$ and $\sigma = 1$, the so-called *standardized* normal distribution. There exist tables for $\Phi(x)$, and the above formula $F(x) = \Phi((x - \mu)/\sigma)$ can be used to compute the distribution function of a normal distribution with arbitrary parameters μ and σ. ☐

One can show that every distribution function $F(x)$ can be represented in the form $F(x) = \lambda_1 F_d(x) + \lambda_2 F_a(x) + \lambda_3 F_s(x)$ where $\lambda_1 + \lambda_2 + \lambda_3 = 1$ and $0 \leqslant \lambda_1, \lambda_2, \lambda_3 \leqslant 1$. Here $F_d(x)$ is a discrete step function, $F_a(x)$ is absolutely continuous, i.e. can be represented as in (6), and $F_s(x)$ is continuous but not absolutely continuous. As already mentioned, $\lambda_3 = 0$ for those distributions which one meets in applications and usually we have either $\lambda_2 = 0$ (for discrete random variables) or $\lambda_1 = 0$ (for continuous random variables).

Sometimes it is convenient to allow a random variable X to take on the value $+\infty$, i.e. $P[X = +\infty] > 0$. If we again set $F(x) = P[X \leqslant x]$ then $\lim_{x \to \infty} F(x) = F(+\infty) = P[X < +\infty] = 1 - P[X = +\infty] < 1$. A distribution function $F(x)$ with $F(+\infty) < 1$ is called *deficient* or *incomplete*. Just as an ordinary distribution function determines a random variable, so does a deficient distribution function $F(x)$ determine a random variable X with $P[X = +\infty] = 1 - F(+\infty) > 0$.

Frequently one does not just consider a single random variable X but several random variables $X_1, X_2, \ldots, X_n, n > 1$, on a probability space $(\Omega, \mathscr{A}, P)$. Just as for a single random variable, we write $[X_1 \leqslant x_1, X_2 \leqslant x_2, \ldots, X_n \leqslant x_n]$ for the event $\{\omega \colon X_1(\omega) \leqslant x_1, X_2(\omega) \leqslant x_2, \ldots, X_n(\omega) \leqslant x_n\}$. The function $F(x_1, x_2, \ldots, x_n)$ which is defined for $-\infty < x_i < \infty, i = 1, \ldots, n$, by

$$F(x_1, \ldots, x_n) = P[X_1 \leqslant x_1, \ldots, X_n \leqslant x_n] \tag{9}$$

is then called the (n-dimensional) *distribution function* in $X_1, \ldots, X_n$. Just as in the one-dimensional case, the n-tuple of random variables defines a probability measure on the Borel algebra of $\mathbb{R}^n$.

If one lets $x_2, \ldots, x_n$ tend to ∞ in (9) one obtains a function $F_1(x_1) = F(x_1, \infty, \ldots, \infty)$ of x_1, the *marginal distribution* with respect to x_1; $F_1(x_1)$ is nothing but the *distribution function* of X_1. If one lets all components x_i, except for x_j, tend to ∞ in (9), the result is once again the marginal distribution with respect to x_j, or the distribution function of X_j. More generally, if one lets $x_{m+1}, \ldots, x_n$ tend to ∞ then $F_{1 \ldots m}(x_1, \ldots, x_m) = F(x_1, \ldots, x_m, \infty, \ldots, \infty)$ is the marginal distribution with respect to $x_1, \ldots, x_m$ or the m-dimensional distribution function of $X_1, \ldots, X_m$. An analogous definition is used for any subset of the coordinates $(x_1, \ldots, x_n)$.

The n random variables $X_1, \ldots, X_n$ are mutually *independent* if the n-dimensional distribution function is the product of the one-dimensional distribution functions (marginal distributions)

$$F(x_1, \ldots, x_n) = F_1(x_1) \cdot \ldots \cdot F_n(x_n). \tag{10}$$

In particular the events $[X_1 \leqslant x_1], [X_2 \leqslant x_2], \ldots, [X_n \leqslant x_n]$ are mutually independent (see (1.1.21)).

Examples

(h) *Multi-dimensional discrete random variables.* Let the X_i, $i = 1, 2, \ldots, n$, be discrete random variables, where X_i could take on possible values $x_i^{(h)}$, $h = 1, 2, \ldots$. Exactly as in the one-dimensional case, the distribution of $X_1, \ldots, X_n$ is determined by the probabilities

$$P[X_1 = x_1^{(h_1)}, \ldots, X_n = x_n^{(h_n)}]$$

for all possible combinations of the possible values $x_i^{(h)}$. The marginal distribution with respect to x_1 is determined by

$$P[X_1 = x_1^{(h)}] = \sum_{h_2, \ldots, h_m} P[X_1 = x_1^{(h_1)}, X_2 = x_2^{(h_2)}, \ldots, X_n = x_n^{(h_n)}]$$

and the marginal distributions with respect to x_i are given analogously. The discrete random variables $X_1, \ldots, X_n$ are *independent* if and only if

$$P[X_1 = x_1^{(h_1)}, \ldots, X_n = x_n^{(h_n)}] = p_1^{(h_1)} \ldots p_n^{(h_n)}.$$

From (1.1.20) we get

$$P[X_1 = x_1^{(h_1)}, \ldots, X_n = x_n^{(h_n)}]$$
$$= P[X_1 = x_1^{(h_1)}] P[X_2 = x_2^{(h_2)} | X_1 = x_1^{(h_1)}]$$
$$\ldots P[X_n = x_n^{(h_n)} | X_1 = x_1^{(h_1)}, \ldots, X_{n-1} = x]. \qquad \square$$

(i) *Multi-dimensional continuous random variables.* If there exists a non-negative function $f(y_1, \ldots, y_n)$ such that

$$F(x_1, \ldots, x_n) = \int_{-\infty}^{x_1} \cdots \int_{-\infty}^{x_n} f(y_1, \ldots, y_n) dy_1 \ldots dy_n,$$

one calls the random variables $X_1, \ldots, X_n$ with distribution function $F(x_1, \ldots, x_n)$ *continuous*. The function $f(y_1, \ldots, y_n)$ is the *n-dimensional density function*. For the marginal distribution with respect to x_1 we obtain

$$F_1(x_1) = F(x_1, \infty, \ldots, \infty)$$

$$= \int_{-\infty}^{x_1} \int_{-\infty}^{\infty} \cdots \int_{-\infty}^{\infty} f(y_1, \ldots, y_n) dy_1 dy_2 \ldots dy_n.$$

Hence

$$F_1(x_1) = \int_{-\infty}^{x_1} f_1(y_1) \, dy_1,$$

where

$$f_1(y_1) = \int_{-\infty}^{\infty} \cdots \int_{-\infty}^{\infty} f(y_1, y_2, \ldots, y_n) \, dy_2 \ldots dy_n.$$

The marginal distribution $F_1(x_1)$ has the density function $f_1(y_1)$, and X_1, considered on its own, is continuous. The same is true for the marginal distribution with respect to x_i. We conclude that the random variables $X_1, \ldots, X_n$ are independent if and only if

$$f(y_1, \ldots, y_n) = f_1(y_1) \ldots f_n(y_n). \qquad \square$$

If $g(x_1, \ldots, x_n)$ is a real valued function defined on $\mathbb{R}^n$ such that for every y with $-\infty < y < \infty$, the set of all n-tuples $(x_1, \ldots, x_n)$ with $g(x_1, \ldots, x_n) \leqslant y$, belongs to the Borel algebra of $\mathbb{R}^n$, then $g(x_1, \ldots, x_n)$ is a random variable. Furthermore, if $X_1(\omega), \ldots, X_n(\omega)$ are random variables on a probability space $(\Omega, \mathscr{A}, P)$, then $Y(\omega) = g(X_1(\omega), \ldots, X_n(\omega))$, or more briefly $Y = g(X_1, \ldots, X_n)$, is also a random variable on the same probability space.

If T is an arbitrary set and $\{X_t, t \in T\}$ is a family of random variables on a probability space $(\Omega, \mathscr{A}, P)$ indexed by t from the set T, then $\{X_t, t \in T\}$ is called a *stochastic process*. In the following chapters T is either equal to the set of *natural* numbers, $T = \{0, 1, 2, \ldots\}$, or equal to the set of non-negative *real* numbers $0 \leqslant t < \infty$. In the first case one speaks of a stochastic process with *discrete* parameter, and in the second case of a process with *continuous* parameter.

$X_t(\omega)$, $\omega \in \Omega$, $t \in T$, can be considered as a function of two variables ω and t. If one fixes t, then $X_t(\omega)$, as a function of ω, is a random variable. If one fixes ω, then $X_t(\omega)$, as a function of t, is a *sample function* of the stochastic process.

If $t_0 < t_1 < \ldots < t_n$ is a finite set of indices in T then one can consider the associated *finite-dimensional* distribution

$$P[X_{t_0} \leqslant x_0, X_{t_1} \leqslant x_1, \ldots, X_{t_n} \leqslant x_n].$$

The collection of all finite-dimensional distributions determines the probabilistic behaviour of the stochastic process. In this generality only very few statements about stochastic processes can be made. In the following chapters, however, additional assumptions of various kinds will be introduced which will lead to specific results.

1.3. Integration and moments

The integration of random variables plays a fundamental role in probability theory. When integrating random variables one applies the Lebesgue–Stieltjes integral whose basic properties are given here without proofs.

Let $(\Omega, \mathscr{A}, P)$ be a probability space and $X(\omega)$ a random variable defined on it. If $X(\omega)$ takes on only *finitely* many different values $x_1, \dots, x_n$ then X is called a *simple* random variable. In this case one defines

$$\int X(\omega) \, dP(\omega) = \sum_{i=1}^{n} x_i P[X = x_i]. \tag{1}$$

More generally, if A is any event in $\mathscr{A}$ and $A_i = [X = x_i] \in \mathscr{A}$ then one defines

$$\int_A X(\omega) \, dP(\omega) = \sum_{i=1}^{n} x_i P(A \cap A_i). \tag{2}$$

Secondly, if $X(\omega)$ is *bounded* (but taking infinitely many values) then one can show that there exists a sequence of *simple* random variables $X_1(\omega)$, $X_2(\omega), \dots$ such that $\lim_{i \to \infty} X_i(\omega) = X(\omega)$ uniformly for all $\omega \in \mathscr{A}$ and such that, for every $A \in \mathscr{A}$,

$$\lim_{i \to \infty} \int_A X_i(\omega) \, dP(\omega) \tag{3}$$

exists and is independent of the particular convergent sequence $X_1(\omega)$, $X_2(\omega), \dots \to X(\omega)$. The limit (3) is denoted by

$$\int_A X(\omega) \, dP(\omega). \tag{4}$$

Thirdly, if $X(\omega)$ is unbounded but *non-negative* ($X(\omega) \geqslant 0$ for all $\omega \in \Omega$) then $X(\omega)$ is called *integrable* if there exists a sequence of *bounded* random variables $X_1(\omega) \leqslant X_2(\omega) \leqslant \dots$ such that

$$\lim_{i \to \infty} X_i(\omega) = X(\omega) \tag{5}$$

for all $\omega \in \Omega$ and

$$\lim_{i \to \infty} \int X_i(\omega) \, dP(\omega) < \infty.$$

In this case one can show that

$$\lim_{i\to\infty} \int_A X_i(\omega)\,dP(\omega) < \infty \tag{6}$$

for all $A \in \mathscr{A}$ and that the limits (6) are independent of the particular sequence satisfying (5). Again the limits (6) are denoted by

$$\int_A X(\omega)\,dP(\omega). \tag{7}$$

Finally, if $X(\omega)$ is an *arbitrary* random variable then $X(\omega)$ is called *integrable* if $X(\omega)$ can be written as a difference $X'(\omega) - X''(\omega)$ of two *non-negative integrable* random variables. In this case one defines

$$\int_A X(\omega)\,dP(\omega) = \int_A X'(\omega)\,dP(\omega) - \int_A X''(\omega)\,dP(\omega). \tag{8}$$

It can be shown that the difference (8) is independent of the special choice of $X'(\omega), X''(\omega)$. The integrals (1), (2), (4), (7) and (8) are *Lebesgue integrals* of the random variables $X(\omega)$.

If $X(x) = g(x)$ is a random variable which is defined on the sample space $(\mathbb{R}, \mathscr{B}, P)$ and if the probability measure P determines the distribution function $F(x)$ on $\mathbb{R}$, then we write for the Lebesgue integral

$$\int_B X(x)\,dP(x) = \int_B g(x)\,dF(x), \quad B \in \mathscr{B}. \tag{9}$$

The second integral in (9) is called a *Lebesgue–Stieltjes integral.* If $F(x)$ is a *discrete* distribution with jumps p_i at the points $x_i, i = 1, 2, \ldots$, then

$$\int_B g(x)\,dF(x) = \sum_{x_i \in B} g(x_i)p_i. \tag{10}$$

If B is equal to one of the intervals $(a, b]$, $[a, b]$, (a, b) or $[a, b)$, we write

$$\int_a^b g(x)\,dF(x), \quad \int_{a-0}^b g(x)\,dF(x), \quad \int_a^{b-0} g(x)\,dF(x), \quad \int_{a-0}^{b-0} g(x)\,dF(x)$$

respectively, for the integral (9).

Between these integrals we have the following relations

$$\int_{a-0}^b g(x)\,dF(x) = g(a)(F(a) - F(a-0)) + \int_a^b g(x)\,dF(x),$$

$$\int_a^{b-0} g(x)\,dF(x) = \int_a^b g(x)\,dF(x) - g(b)(F(b) - F(b-0)),$$

$$\int_{a-0}^{b-0} g(x)\,dF(x) = g(a)\,(F(a) - F(a-0))$$

$$- g(b)\,(F(b) - F(b-0)) + \int_a^b g(x)\,dF(x). \quad (11)$$

The reader can check this for an example of a discrete distribution function $F(x)$.

In the case of a *continuous* distribution function $F(x)$ the values of the four integrals obviously coincide. If $f(y)$ is the density function of $F(x)$ and if $g(x)$ is (piecewise) continuous then one can show that

$$\int_a^b g(x)\,dF(x) = \int_a^b g(x)f(x)\,dx, \quad (12)$$

where the expression on the right is an ordinary *Riemann integral*. For the two most important cases of discrete and continuous distributions one has therefore in (10) and (12) explicitly computable formulae for the integration of random variables.

$dF(x) = P[x \leqslant X \leqslant x + dx]$ is called the *probability element* of x. For discrete distribution functions with jumps p_i at the points x_i, $dF(x) = 0$ for $x \neq x_i$ and $dF(x_i) = p_i$. For continuous distribution functions with density function $f(x)$, $dF(x) = f(x)\,dx$. This emerges if one compares the integration formulae (10) and (12).

If X is a random variable on a probability space $(\Omega, \mathscr{A}, P)$, $F(x)$ the distribution function of X and $g(x)$ a random variable with respect to the Borel algebra of $\mathbb{R}$, then, as shown in Section 1.2, $g(X)$ is a random variable on $(\Omega, \mathscr{A}, P)$. If B is an event in the Borel algebra $\mathscr{B}$ and $A = [X \in B]$, one can show that

$$\int_A g(X(\omega))\,dP(\omega) = \int_B g(x)\,dF(x). \quad (13)$$

To conclude this short introduction to Lebesgue–Stieltjes integration theory for random variables, let us collect together some useful properties of integrals.

If $X(\omega) = c$ for every $\omega \in \Omega$ (degenerate random variable, Example $(1.2.a)$), then

$$\int_A X(\omega)\,dP(\omega) = cP(A). \quad (14)$$

If $a \leqslant X(\omega) \leqslant b$ for every $\omega \in \Omega$, then

$$aP(A) \leqslant \int_A X(\omega)\,dP(\omega) \leqslant bP(A). \quad (15)$$

If $X_1(\omega), X_2(\omega)$ are both integrable random variables, then $aX_1(\omega) + bX_2(\omega)$ is also an integrable random variable and

$$\int_A (aX_1(\omega) + bX_2(\omega))dP(\omega) = a \int_A X_1(\omega)dP(\omega) + b \int_A X_2(\omega)dP(\omega).$$

(16)

If $X_1(\omega) \leqslant X_2(\omega)$ holds for every $\omega \in \Omega$ then

$$\int_A X_1(\omega)dP(\omega) \leqslant \int_A X_2(\omega)dP(\omega).$$

(17)

$X(\omega)$ is integrable if and only if $|X(\omega)|$ is integrable and

$$\left| \int_A X(\omega)dP(\omega) \right| \leqslant \int_A |X(\omega)| dP(\omega).$$

(18)

These results are frequently used in the following chapters without explicit reference and hold, of course, also for integration on the probability space $(\mathbb{R}, \mathscr{B}, P)$, i.e. for integration with respect to $dF(x)$, where $F(x)$ is the distribution function determined by P. It follows from (14) that, in particular,

$$\int_{-\infty}^{y} dF(x) = F(y), \quad \int_{a}^{b} dF(x) = F(b) - F(a)$$

(19)

(see also (11)). In the left hand integral the lower bound can be replaced by 0 if $F(x) = 0$ for $x \leqslant 0$, i.e. if the associated random variable can only take on positive values.

The integral of an integrable random variable X over the entire sample space is called the *expected value* of the random variable and denoted by $E[X]$. Using (13) we note that

$$E[X] = \int X(\omega)dP(\omega) = \int_{-\infty}^{\infty} x dF(x),$$

(20)

where $F(x)$ is the distribution function of X. Equations (10) and (12) imply for *discrete* and *continuous* random variables that

$$E[X] = \sum_i x_i p_i, \quad E[X] = \int_{-\infty}^{\infty} x f(x)dx,$$

(21)

if the discrete random variable X takes on the possible values x_i with probabilities $p_i, i = 1, 2, \ldots$, and if the continuous random variable X possesses the density function $f(x)$.

The integration properties (14) to (18) have their analogues for expectation values: for a *degenerate random variable*, $X(\omega) = c$ for every $\omega \in \Omega$,

$$E[X] = c.$$

(22)

If $a \leqslant X(\omega) \leqslant b$ for every $\omega \in \Omega$ then

$$a \leqslant E[X] \leqslant b. \tag{23}$$

Equation (16) yields

$$E[aX_1 + bX_2] = aE[X_1] + bE[X_2]. \tag{24}$$

With $a = b = 1$ it follows from this, in particular, that the expected value of a sum of random variables is equal to the sum of the expected values, irrespective of whether the random variables are independent or not:

$$E\left[\sum_{i=1}^{n} X_i\right] = \sum_{i=1}^{n} E[X_i]. \tag{25}$$

If $X_1(\omega) \leqslant X_2(\omega)$ for every $\omega \in \Omega$ then

$$E[X_1] \leqslant E[X_2], \tag{26}$$

and, finally,

$$|E[X_1]| \leqslant E[|X_1|]. \tag{27}$$

The expected values of the random variables $g(X) = X^r, r = 1, 2, \ldots,$ are called rth *moments* of the random variable X. Of particular importance is the second moment of $X - E[X], E[(X - E[X])^2]$, the *variance* of the random variable X. From (22) and (24) it follows that $E[X - E[X]] = 0$, which together with (24), implies

$$E[(X - E[X])^2] = E[X^2 - 2XE[X] + (E[X])^2] = E[X^2] - (E[X])^2. \tag{28}$$

According to (20) and (21) we have in the general as well as in the discrete and continuous cases

$$E[(X - E[X])^2] = \int_{-\infty}^{\infty} (x - E[X])^2 \, dF(x)$$

$$= \int_{-\infty}^{\infty} x^2 \, dF(x) - \left(\int_{-\infty}^{\infty} x \, dF(x)\right)^2, \tag{29}$$

$$E[(X - E[X])^2] = \sum_i (x_i - E[X])^2 p_i = \sum_i x_i^2 p_i - \left(\sum_i x_i p_i\right)^2, \tag{30}$$

$$E[(X - E[X])^2] = \int_{-\infty}^{\infty} (x - E[X])^2 f(x) \, dx$$

$$= \int_{-\infty}^{\infty} x^2 f(x) \, dx - \left(\int_{-\infty}^{\infty} x f(x) \, dx\right)^2. \tag{31}$$

If X is a random variable with expected value $\mu = E[X]$, variance $\sigma^2 = E[(X - \mu)^2]$ and distribution function $F(x)$, and if $\lambda > 0$, then

$$\sigma^2 = \int_{\mu-\lambda\sigma}^{\mu+\lambda\sigma} (x-\mu)^2 \, dF(x) + \int_{-\infty}^{\mu-\lambda\sigma} (x-\mu)^2 \, dF(x) + \int_{\mu+\lambda\sigma}^{\infty} (x-\mu)^2 \, dF(x)$$

If one omits the first integral term and replaces x in the two remaining integrands by the upper and lower bound, respectively, then

$$\sigma^2 \geqslant \int_{-\infty}^{\mu-\lambda\sigma} (\lambda\sigma)^2 \, dF(x) + \int_{\mu+\lambda\sigma}^{\infty} (\lambda\sigma)^2 \, dF(x) = (\lambda\sigma)^2 P[|X-\mu| \geqslant \lambda\sigma],$$

and from this we obtain *Tchebychev's inequality*

$$P[|X-\mu| \geqslant \lambda\sigma] \leqslant 1/\lambda^2. \tag{32}$$

This shows that the values of the random variable lie with high probability in a neighbourhood of the expected value μ. The square root of the variance, the *standard deviation*, is, according to (32), a measure for the expected deviation of the values of the random variable from its expected value. Expected values, standard deviations and variances represent therefore a simple and rough numerical classification of random variables and their distribution functions, respectively.

Examples

(a) *Degenerate random variables* (see Example (1.2.*a*)). For a degenerate random variable, $X = c$, it has already been observed in (22) that $E[X] = c$. $(X-E[X])^2 = 0$ is also a degenerate random variable and the variance of X is therefore $E[(X-E[X])^2] = 0$. The converse also holds: if a random variable X has variance 0 then it is degenerate. More precisely, $X(\omega) = c$ has to hold in this case for all $\omega \in \Omega$ except for certain ω which belong to an event A with probability 0. □

(b) *Indicator functions.* Let $(\Omega, \mathscr{A}, P)$ be a probability space and A an event in $\mathscr{A}$. The function $I_A(\omega) = 1$ for $\omega \in A$, $I_A(\omega) = 0$ for $\omega \notin A$, is a random variable and is called the *indicator function* of A. It follows from (1) that $E[I_A] = P[I_A = 1] = P(A)$. If $F(x)$ is a distribution function and if A runs successively through the intervals $(a, b]$, $[a, b]$, (a, b) and $[a, b)$, then

$$E[I_{(a,b]}] = \int_a^b dF(x) = F(b) - F(a),$$

$$E[I_{[a,b]}] = \int_{a-0}^b dF(x) = F(b) - F(a-0),$$

$$E[I_{(a,b)}] = \int_a^{b-0} dF(x) = F(b-0) - F(a),$$

$$E[I_{[a,b)}] = \int_{a-0}^{b-0} dF(x) = F(b-0) - F(a-0). \qquad \square$$

(c) *Bernoulli variables.* If X is a Bernoulli variable (see Example (1.2.*b*)), then $E[X] = 0(1-p) + 1p = p$ and $E[(X - E[X])^2] = (0-p)^2(1-p) + (1-p)^2 p = p(1-p)$. □

(d) *Geometric distributions.* If X is a discrete random variable with a geometric distribution (see Example (1.2.*c*)), then the expected value $E[X] = (1-p)(p + 2p^2 + 3p^3 + \ldots) = (1-p)p(1 + 2p + 3p^2 + \ldots)$. The series in this formula is equal to the derivative of the series $p + p^2 + p^3 + \ldots = p/(1-p)$ with respect to p. Therefore

$$E[X] = p(1-p)\frac{\mathrm{d}}{\mathrm{d}p}\frac{p}{1-p} = \frac{p}{1-p}.$$ □

(e) *Poisson distribution.* The expected value of a discrete random variable with a Poisson distribution (see Example (1.2.*d*)) is equal to

$$E[X] = \sum_{i=1}^{\infty} i\frac{\lambda^i e^{-\lambda}}{i!} = \lambda e^{-\lambda}\sum_{i=0}^{\infty}\frac{\lambda^i}{i!} = \lambda e^{-\lambda}e^{\lambda} = \lambda.$$

The second moment satisfies

$$E[X^2] = \sum_{i=1}^{\infty} i^2\frac{\lambda^i e^{-\lambda}}{i!} = \lambda\sum_{i=1}^{\infty} i\frac{\lambda^{i-1}e^{-\lambda}}{(i-1)!}$$

$$= \lambda\sum_{i=1}^{\infty}(i-1)\frac{\lambda^{i-1}e^{-\lambda}}{(i-1)!} + \lambda\sum_{i=1}^{\infty}\frac{\lambda^{i-1}e^{-\lambda}}{(i-1)!} = \lambda^2 + \lambda.$$

According to (28) the variance of X is equal to $(\lambda^2 + \lambda) - \lambda^2 = \lambda$. The variance of a Poisson distribution is thus equal to its expected value. □

(f) *Uniform distribution.* If the random variable X has a uniform distribution in the interval $[a, b]$ (see Example (1.2.*e*)) then

$$E[X] = \int_a^b \frac{x}{b-a}\,\mathrm{d}x = \frac{b+a}{2}, \quad E[X^2] = \int_a^b \frac{x^2}{b-a}\,\mathrm{d}x = \frac{b^2 + ab + a^2}{3}.$$

According to (28) it therefore follows for the uniform distribution in the interval $[a, b]$ that $E[(X - E[X])^2] = (b-a)^2/12$. □

(g) *Exponential distributions.* For an exponentially distributed random variable X (see Example (1.2.*f*)) integration by parts yields

$$E[X] = \int_0^{\infty} \lambda x e^{-\lambda x}\,\mathrm{d}x = [-x e^{-\lambda x}]_0^{\infty} + \int_0^{\infty} e^{-\lambda x}\,\mathrm{d}x = \frac{1}{\lambda}.$$

$$E[X^2] = \int_0^{\infty} \lambda x^2 e^{-\lambda x}\,\mathrm{d}x = [-x^2 e^{-\lambda x}]_0^{\infty} + 2\int_0^{\infty} x e^{-\lambda x}\,\mathrm{d}x = \frac{2}{\lambda^2}.$$

According to (28) the variance satisfies $E[(X - E[X])^2] = 1/\lambda^2$. With exponential distributions the standard deviation is equal to the expected value. □

(*h*) *Normal distributions.* The expected value of a normally distributed random variable with the parameters μ and σ (see Example (1.2.*g*)) is equal to

$$E[X] = \frac{1}{\sqrt{(2\pi)}\,\sigma} \int_{-\infty}^{\infty} x e^{-(x-\mu)^2/2\sigma^2}\,dx.$$

If one uses the substitution $z = (x - \mu)/\sigma$ one finds that

$$E[X] = \frac{\sigma}{\sqrt{(2\pi)}} \int_{-\infty}^{\infty} z e^{-z^2/2}\,dz + \frac{\mu}{\sqrt{(2\pi)}} \int_{-\infty}^{\infty} e^{-z^2/2}\,dz.$$

The first integral vanishes. The second integral has been computed in Example (1.2.*g*) and is equal to $\sqrt{(2\pi)}$. Therefore $E[X] = \mu$.

The variance is equal to

$$E[(X - \mu)^2] = \frac{1}{\sqrt{(2\pi)}\,\sigma} \int_{-\infty}^{\infty} (x - \mu)^2 e^{-(x-\mu)^2/2\sigma^2}\,dx.$$

The same substitution $z = (x - \mu)/\sigma$ transforms this integral into

$$E[(X - \mu)^2] = \frac{\sigma^2}{\sqrt{(2\pi)}} \int_{-\infty}^{\infty} z^2 e^{-z^2/2}\,dz.$$

Through integration by parts one finds

$$\int_{-\infty}^{\infty} z^2 e^{-z^2/2}\,dz = [-z e^{-z^2/2}]_{-\infty}^{\infty} + \int_{-\infty}^{\infty} e^{-z^2/2}\,dz = \sqrt{(2\pi)}.$$

The variance is therefore equal to σ^2. The parameters μ and σ^2 of a normal distribution are therefore precisely its expected value and variance. □

If $(X_1, X_2, \ldots, X_n)$ is an n-dimensional random vector with the n-dimensional distribution function $F(x_1, x_2, \ldots, x_n)$ (see (1.2.9)), the expected values $E[X_i]$ of the random variables X_i, $i = 1, 2, \ldots, n$, are equal to

$$E[X_i] = \int_{-\infty}^{\infty} x\,dF_i(x),$$

where $F_i(x)$ is the marginal distribution of $F(x_1, x_2, \ldots, x_n)$ with respect to the ith component. Of particular significance are the *mixed second moments*

$$E[X_i X_j] = \int_{-\infty}^{+\infty} \cdots \int_{-\infty}^{+\infty} x_i x_j\,dF(x_1, x_2, \ldots, x_n), \quad i, j = 1, 2, \ldots, n. \tag{33}$$

If the random variables $X_1, X_2, \ldots, X_n$ are *independent* one can show that

$$E[X_i X_j] = E[X_i]E[X_j], \quad i \neq j. \tag{34}$$

The expected value of the product is thus equal to the product of the expected values.

The mixed second moments of the random variables $X_1 - E[X_1], \ldots,$ $X_n - E[X_n]$, i.e. the expected values $\sigma_{ij} = E[(X_i - E[X_i])(X_j - E[X_j])]$, are called *covariances* of the random variables $X_1, X_2, \ldots, X_n$. The covariance σ_{ii} is obviously equal to the variance of X_i. If $X_1, X_2, \ldots, X_n$ are *independent*, then it follows from (34) that $\sigma_{ij} = E[X_i - E[X_i]]E[X_j - E[X_j]] = 0$ for $i \neq j$. For independent random variables the covariances vanish. The converse is not necessarily true: if the covariance of two random variables vanishes then these two random variables do not necessarily have to be independent.

If X and Y are two arbitrary random variables and if λ is a real number then $E[(X + \lambda Y)^2] = E[X^2 + 2\lambda XY + \lambda^2 Y^2] = E[X^2] + 2\lambda E[XY] + \lambda^2 E[Y^2] \geqslant 0$ for all λ. Therefore the discriminant of this quadratic function in λ cannot be positive: $4(E[XY])^2 - 4E[X^2]E[Y^2] \leqslant 0$. From this we obtain the *Schwarz inequality*

$$(E[XY])^2 \leqslant E[X^2]E[Y^2]. \tag{35}$$

Applying (35) to $X = X_i - E[X_i]$, $Y = X_j - E[X_j]$, we conclude $\sigma_{ij}^2 \leqslant \sigma_{ii}\sigma_{jj}$.

To end this section let us prove some useful formulae for *integration by parts* for the Lebesgue–Stieltjes integral. If $g(x)$ is a bounded function with continuous derivative $g'(x)$ then

$$\int_a^b g(x)\,\mathrm{d}F(x) = g(b)F(b) - g(a)F(a) - \int_a^b g'(x)F(x)\,\mathrm{d}x. \tag{36}$$

This is equivalent to

$$\int_a^b (g(b) - g(x))\,\mathrm{d}F(x) - \int_a^b g'(x)(F(x) - F(a))\,\mathrm{d}x = 0. \tag{37}$$

If $|g'(x)| \leqslant C$, then it follows that $|g(b) - g(x)| \leqslant C(b - a)$ for $a \leqslant x \leqslant b$ and the absolute value of the left hand side of (37) is smaller than $2C(b - a) \times (F(b) - F(a))$. If one subdivides the interval $(a, b]$ into equidistant points $a = x_0 < x_1 < \ldots < x_n = b$, where $x_k - x_{k-1} = h$, then the left hand side of (37) is also equal to

$$\sum_{k=1}^n \int_{x_{k-1}}^{x_k} (g(x_k) - g(x))\,\mathrm{d}F(x) - \int_{x_{k-1}}^{x_k} g'(x)(F(x) - F(x_{k-1}))\,\mathrm{d}x.$$

The above estimate for the left hand side of (37) holds for every single term of this sum, i.e. the absolute value of every term is smaller than $2Ch(F(x_k) - F(x_{k-1}))$. From this it follows that the absolute value of the left hand side of (37) has to be smaller than $2Ch(F(b) - F(a))$. Since h is arbitrary the left hand side of (37) must indeed vanish and (36) is thus proved.

If one takes $a = 0$ and $g(x) = x$ in (36), then it follows that

$$\int_0^b x\mathrm{d}F(x) = -b(1-F(b)) + \int_0^b (1-F(x))\mathrm{d}x.$$

If the left hand integral converges for $b \to \infty$, then

$$\int_b^\infty x\mathrm{d}F(x) \geqslant b(1-F(b)) \to 0 \quad \text{as } b \to \infty$$

implies

$$\int_0^\infty x\mathrm{d}F(x) = \int_0^\infty (1-F(x))\mathrm{d}x. \tag{38}$$

The result is a new formula for the *expected value of a positive random variable X* with a distribution function $F(x)$, where $F(x) = 0$ for $x \leqslant 0$.

1.4. Sums of random variables

Let $X_1, X_2, \ldots, X_n$ be arbitrary, not necessarily independent, random variables. The *expected value* of $X_1 + X_2 + \ldots + X_n$ is, according to (1.3.25), equal to the sum of the expected values of $X_1, X_2, \ldots, X_n$

$$E\left[\sum_{i=1}^n X_i\right] = \sum_{i=1}^n E[X_i]. \tag{1}$$

Let $\mu_i = E[X_i], i = 1, 2, \ldots, n$, be the expected values of the random variables X_i. One obtains for the *variance* of the sum $X_1 + X_2 + \ldots + X_n$

$$E\left[\left(\sum_{i=1}^n (X_i - \mu_i)\right)^2\right] = E\left[\sum_{i,j=1}^n (X_i - \mu_i)(X_j - \mu_j)\right]$$

$$= \sum_{i=1}^n E[(X_i - \mu_i)^2] + \sum_{\substack{i,j=1 \\ i \neq j}}^n E[(X_i - \mu_i)(X_j - \mu_j)]$$

$$= \sum_{i=1}^n \sigma_i^2 + \sum_{\substack{i,j=1 \\ i \neq j}}^n \sigma_{ij}. \tag{2}$$

Here σ_i^2 is the variance of X_i, and σ_{ij} is the *covariance* of X_i and X_j (see Section 1.3). If $X_i, i = 1, 2, \ldots, n$, are *independent* random variables then $\sigma_{ij} = 0$ for $i \neq j$ and it follows from (2) that

$$E\left[\left(\sum_{i=1}^n (X_i - \mu_i)\right)^2\right] = \sum_{i=1}^n \sigma_i^2. \tag{3}$$

The variance of a sum of *independent* random variables X_i is equal to the sum of the variances of the X_i.

Next we shall determine the *distribution function* of the sum $X_1 + X_2$ of two *independent* random variables X_1 and X_2. Let the distribution functions of X_1 and X_2 be $F_1(x)$ and $F_2(x)$. Fix $\epsilon > 0$. If the event $[X_1 + X_2 \leqslant y]$ occurs then one of the events $[n\epsilon < X_1 \leqslant (n+1)\epsilon, X_2 \leqslant y - n\epsilon] = [n\epsilon < X_1 \leqslant (n+1)\epsilon] \cap [X_2 \leqslant y - n\epsilon]$ has to occur for some $n = 0, \pm 1, \pm 2, \ldots$, i.e.

$$[X_1 + X_2 \leqslant y] \subset \bigcup_n [n\epsilon < X_1 \leqslant (n+1)\epsilon, X_2 \leqslant y - n\epsilon].$$

Since the events in the union on the right are disjoint,

$$P[X_1 + X_2 \leqslant y] \leqslant \sum_n P[n\epsilon < X_1 \leqslant (n+1)\epsilon, X_2 \leqslant y - n\epsilon]. \qquad (4)$$

From the independence of X_1 and X_2 it follows, furthermore, that

$$P[n\epsilon < X_1 \leqslant (n+1)\epsilon, X_2 \leqslant y - n\epsilon]$$
$$= P[n\epsilon < X_1 \leqslant (n+1)\epsilon] P[X_2 \leqslant y - n\epsilon]$$
$$= F_2(y - n\epsilon)(F_1((n+1)\epsilon) - F_1(n\epsilon))$$

and therefore

$$P[X_1 + X_2 \leqslant y] \leqslant \sum_n F_2(y - n\epsilon)(F_1((n+1)\epsilon) - F_1(n\epsilon)). \qquad (5)$$

On the right hand side of (5) we have

$$\int_{-\infty}^{\infty} F_{2\epsilon}(x) \, dF_1(x),$$

which has as its integrand the step function $F_{2\epsilon}(x)$ which takes the value $F_2(y - n\epsilon)$ in the interval $n\epsilon < x \leqslant (n+1)\epsilon$. Also, $F_{2\epsilon}(x) \leqslant F_2(y + \epsilon - x)$ and hence (by (1.3.17))

$$P[X_1 + X_2 \leqslant y] \leqslant \int_{-\infty}^{\infty} F_2(y + \epsilon - x) \, dF_1(x). \qquad (6)$$

Conversely, if one of the events $[n\epsilon < X_1 \leqslant (n+1)\epsilon, X_2 \leqslant y - (n+1)\epsilon]$, $n = 0, \pm 1, \pm 2, \ldots$, occurs, the event $[X_1 + X_2 \leqslant y]$ also occurs, i.e.

$$[X_1 + X_2 \leqslant y] \supset \bigcup_n [n\epsilon < X_1 \leqslant (n+1)\epsilon, X_2 \leqslant y - (n+1)\epsilon].$$

An analogous consideration leads to the inequality

$$P[X_1 + X_2 \leqslant y] \geqslant \int_{-\infty}^{\infty} F_2(y - x - \epsilon) \, dF_1(x). \qquad (7)$$

Since $\epsilon > 0$ is arbitrary, (6) and (7) imply

$$P[X_1 + X_2 \leqslant y] = \int_{-\infty}^{\infty} F_2(y-x)\mathrm{d}F_1(x). \tag{8}$$

The distribution function of X_1 and X_2 has thus been computed.

Briefly, (8) can be derived as follows: $X_1 + X_2 \leqslant y$ holds if and only if $x < X_1 \leqslant x + \mathrm{d}x$ and $X_2 \leqslant y - x$ for some x, $-\infty < x < +\infty$. Since $P[x < X_1 \leqslant x + \mathrm{d}x] = \mathrm{d}F_1(x)$, these events have the probability $F_2(y-x)\mathrm{d}F_1(x)$. The probability for $X_1 + X_2 \leqslant y$ is thus obtained by summing these probabilities over all x, i.e. by (8). Such shortened argumentation will be used in the following chapters for similar situations.

The integral in (8) is called the *convolution* of the distribution functions F_1 and F_2 and is denoted by $F_2 * F_1$. Since the roles of F_1 and F_2 in (8) can, of course, be interchanged, the *commutative law* $F_1 * F_2 = F_2 * F_1$ holds, i.e.

$$(F_1 * F_2)(y) = \int_{-\infty}^{\infty} F_2(y-x)\mathrm{d}F_1(x) = \int_{-\infty}^{\infty} F_1(y-x)\mathrm{d}F_2(x)$$

$$= (F_2 * F_1)(y). \tag{9}$$

For the sum of three independent random variables X_1, X_2, X_3 with distribution functions F_1, F_2, F_3 the associative law $(X_1 + X_2) + X_3 = X_1 + (X_2 + X_3) = X_1 + X_2 + X_3$ holds, and we obtain the *associative law* for convolutions $(F_1 * F_2) * F_3 = F_1 * (F_2 * F_3) = F_1 * F_2 * F_3$.

The distribution function of the sum of the independent random variables $X_1 + \ldots + X_n$ is given by the n-fold convolution $F_1 * F_2 * \ldots * F_n$. Of particular importance is the case where n random variables $X_1, \ldots, X_n$ possess the same distribution function $F_1(x) = F_2(x) = \ldots = F_n(x) = F(x)$. For the n-fold convolution $F * F * \ldots * F$ one simply writes F^{*n}. All n random variables in this case have the *same* expected value $E[X_i] = \mu$ and the *same* variance $E[(X_i - \mu)^2] = \sigma^2$. Hence the expected value and variance of the sum $X_1 + \ldots + X_n$ are, according to (1) and (3), equal to $n\mu$ and $n\sigma^2$.

If X_1 and X_2 are independent random variables which can only take positive values then the distribution functions of the two random variables satisfy $F_1(x) = F_2(x) = 0$ for all $x \leqslant 0$. The convolution in this case is equal to

$$\int_0^y F_2(y-x)\mathrm{d}F_1(x) = \int_0^y F_1(y-x)\mathrm{d}F_2(x). \tag{10}$$

If X_1, X_2 are *continuous* random variables (see Section 1.2) with density functions $f_1(x), f_2(x)$, then it follows from (8) that

$$P[X_1 + X_2 \leqslant y] = \int_{-\infty}^{\infty} \left(\int_{-\infty}^{y-x} f_2(z)\mathrm{d}z \right) f_1(x)\mathrm{d}x$$

$$= \int_{-\infty}^{\infty} \left(\int_{-\infty}^{y} f_2(v-x)\mathrm{d}v \right) f_1(x)\mathrm{d}x$$

$$= \int_{-\infty}^{y} \left(\int_{-\infty}^{\infty} f_2(v-x)f_1(x)\mathrm{d}x \right) \mathrm{d}v.$$

This implies that $X_1 + X_2$ is also *continuous* and has the density function

$$f(y) = \int_{-\infty}^{\infty} f_2(y - x) f_1(x) \mathrm{d}x. \tag{11}$$

The integral in (11) is called the convolution of density functions.

Examples
(a) *Discrete distributions on the natural numbers.* Let X_1 and X_2 be *discrete* independent random variables which can only take the values of non-negative integers $0, 1, 2, \ldots$. The distribution of the two random variables can therefore be described by the probabilities $p_i^{(1)} = P[X_1 = i]$, $p_i^{(2)} = P[X_2 = i]$. The sum $X_1 + X_2$ can also only take the values $0, 1, 2, \ldots$ and the distribution of $X_1 + X_2$ can again be described by the probabilities $p_i = P[X_1 + X_2 = i]$. The event $[X_1 + X_2 = i]$ occurs when one of the events $[X_1 = 0, X_2 = i]$, $[X_1 = 1, X_2 = i - 1]$, $\ldots$, $[X_1 = i, X_2 = 0]$ occurs. All these events are disjoint and, because of the independence of the two random variables, the event $[X_1 = j, X_2 = i - j]$ has the probability $p_j^{(1)} p_{i-j}^{(2)}$. Hence

$$p_i = \sum_{j=0}^{i} p_{i-j}^{(2)} p_j^{(1)}, \quad i = 0, 1, 2, \ldots.$$

This operation is called *convolution* of the sequence $\{p_i^{(1)}, i = 0, 1, 2, \ldots\}$ and $\{p_i^{(2)}, i = 0, 1, 2, \ldots\}$. □

(b) *Binomial distributions.* Let $X_1, X_2, \ldots, X_n$ be *independent Bernoulli variables* (see Example (1.2.b)). Suppose $X_i = 1$; then this can be interpreted as 'success' at the ith attempt. The sum $X_1 + X_2 + \ldots + X_n$ thus represents the number of 'successes' during n (independent) attempts. What is the distribution of this sum?

As in Example (1.2.b) let $p = P[X_i = 1]$ and $1 - p = P[X_i = 0]$. Furthermore let $p_i^{(n)} = P[X_1 + \ldots + X_n = i]$, $i = 0, 1, \ldots, n$. According to Example (a)

$$p_0^{(2)} = (1 - p)^2, \quad p_1^{(2)} = 2p(1 - p), \quad p_2^{(2)} = p_2.$$

We show by induction on n that

$$p_i^{(n)} = \binom{n}{i} p^i (1 - p)^{n-i}, \quad i = 0, 1, 2, \ldots, n.$$

Here

$$\binom{n}{i} = \frac{n!}{i!(n-i)!}, \quad \binom{n}{0} = \binom{n}{n} = 1$$

are the well-known *binomial coefficients* and the distribution $p_i^{(n)}$ is called the *binomial distribution.*

The assertion holds for $n = 2$. If we assume it holds for n, then it follows from Example (a) that

$$p_i^{(n+1)} = p_i^{(n)}(1-p) + p_{i-1}^{(n)}p = \binom{n}{i}p^i(1-p)^{n+1-i} + \binom{n}{i-1}p^i(1-p)^{n+1-i}.$$

Since

$$\binom{n}{i} + \binom{n}{i-1} = \frac{n!}{i!(n-i)!} + \frac{n!}{(i-1)!(n+1-i)!}$$

$$= \frac{(n+1)!}{i!(n+1-i)!} = \binom{n+1}{i},$$

it follows that the assertion also holds for $n + 1$, and thus for all n.

The binomial distribution has two parameters, n and p. If X_1, X_2 are two *independent* random variables which are both *binomially* distributed, with parameters n and m and with the *same* parameter p, it should be obvious from the above that $X_1 + X_2$ is binomially distributed, with parameters $(n + m)$ and p.

The expected value and variance of the binomial distribution are best determined from the expected value $\mu = p$, and the variance $\sigma^2 = p(1-p)$ of the Bernoulli variable (Example (1.3.c)) with the help of (1) and (3). The expected value of the binomial distribution is thus equal to np and the variance equal to $np(1-p)$. $\square$

(c) *Sums of Poisson-distributed random variables.* Let X_1 and X_2 be *independent* random variables with Poisson distribution with parameters λ_1 and λ_2, respectively (see Example (1.2.d)). Furthermore, let $p_i = P[X_1 + X_2 = i]$. By Example (a)

$$p_i = \sum_{j=0}^{i} \frac{\lambda_1^j \lambda_2^{i-j}}{j!(i-j)!} e^{-(\lambda_1 + \lambda_2)}.$$

By applying the *binomial theorem*:

$$(\lambda_1 + \lambda_2)^i = \sum_{j=0}^{i} \binom{i}{j}\lambda_1^j\lambda_2^{i-j} = \sum_{j=0}^{i} \frac{i!}{j!(i-j)!}\lambda_1^j\lambda_2^{i-j},$$

it can be shown that

$$p_i = \frac{(\lambda_1 + \lambda_2)^i}{i!} e^{-(\lambda_1 + \lambda_2)}.$$

The sum $X_1 + X_2$ is thus again *Poisson distributed* with parameter $\lambda_1 + \lambda_2$. $\square$

The convolution operation (9) is not only defined between distribution functions. If F is a distribution function and g an arbitrary function (more precisely, a $\mathscr{B}$-measurable function; see Section 1.2), the convolution $g * F$ can be defined. Here we shall define this operation only for distributions which are concentrated on the *positive* half-axis, i.e. for all $F(x) = 0$ for $x \leqslant 0$ and for functions $g(x)$ with $g(x) = 0$ for $x \leqslant 0$. In this case the definition is

$$(g*F)(y) = \int_0^y g(y-x)\mathrm{d}F(x). \tag{12}$$

If $g(x)$ is not a distribution function then the convolution $F*g$ has *no* meaning, and in particular the commutative law is not valid for this generalization.

Furthermore, functions $F(x)$ will have to be considered which have every property of a distribution function (see Example $(1.1.j)$), except that $\lim_{x\to\infty}F(x) \neq 1$: it may even happen that $\lim_{x\to\infty}F(x) = \infty$. Just as a distribution function determines a probability measure P on the Borel algebra, a generalized distribution function defines a measure P on the Borel algebra which has every property of a probability measure, except that the entire space does not have measure 1 any more (see Section 1.1, conditions for probability measures). Consequently, the Lebesgue–Stieltjes integral with respect to $\mathrm{d}P(x)$ or $\mathrm{d}F(x)$ can be defined exactly as for a distribution function (see Section 1.3) and the integrals have again the same properties. The convolution (12), in particular, is also defined in this more generalized case.

For the convolution (12) we have two *distributive laws*

$$(g_1 + g_2)*F = \int_0^y (g_1(y-x) + g_2(y-x))\mathrm{d}F(x)$$

$$= \int_0^y g_1(y-x)\mathrm{d}F(x) + \int_0^y g_2(y-x)\mathrm{d}F(x)$$

$$= g_1*F + g_2*F \tag{13}$$

and

$$g*(F_1 + F_2) = \int_0^y g(y-x)(\mathrm{d}F_1(x) + \mathrm{d}F_2(x))$$

$$= \int_0^y g(y-x)\mathrm{d}F_1(x) + \int_0^y g(y-x)\mathrm{d}F_2(x)$$

$$= g*F_1 + g*F_2. \tag{14}$$

Finally, let us consider the arithmetic mean of n *independent* random variables:

$$\bar{X}_n = \frac{1}{n}\sum_{i=1}^n X_i, \tag{15}$$

in particular with respect to their behaviour as $n \to \infty$. The random variables are each to have the *same* distribution function and thus the same expected value $E[X_i] = \mu$ and variance $E[(X_i - \mu)^2] = \sigma^2$. We assume that these moments exist and are *finite*, $\mu < \infty$, $\sigma^2 < \infty$. The expected value of $\bar{X}_n$ is, according to (1), given by

$$E[\bar{X}_n] = \frac{1}{n} E\left[\sum_{i=1}^{n} X_i\right] = \frac{1}{n} \sum_{i=1}^{n} E[X_i] = \mu. \tag{16}$$

The arithmetic mean $\bar{X}_n$ has thus the same expected value μ as have the random variables X_i. For the variance σ_n^2 of the arithmetic mean we obtain by using (3)

$$\sigma_n^2 = E\left[\left(\frac{1}{n} \sum_{i=1}^{n} X_i - \mu\right)^2\right] = \frac{1}{n^2} E\left[\left(\sum_{i=1}^{n} (X_i - \mu)\right)^2\right] = \frac{\sigma^2}{n}. \tag{17}$$

The variance of the arithmetic mean $\bar{X}_n$ tends to zero as $n \to \infty$. Choose $\epsilon > 0$. Then it follows from Tchebychev's inequality (1.3.32), applied to $\bar{X}_n$, that

$$P[|\bar{X}_n - \mu| \geqslant \epsilon] \leqslant \sigma^2/\epsilon^2 n. \tag{18}$$

For every $\epsilon > 0$ we have therefore

$$\lim_{n \to \infty} P[|\bar{X}_n - \mu| \geqslant \epsilon] = 0. \tag{19}$$

For $n \to \infty$, X_n is said to *converge in probability* to the expected value μ. This is the so-called *weak law of large numbers*.

Moreover, if the $X_i, i = 1, 2, \ldots$, are defined on the probability space $(\Omega, \mathscr{A}, P)$, then $\bar{X}_n$ is also a function on the same probability space. It can be shown that the event $\{\omega : \lim_{n \to \infty} \bar{X}_n(\omega) = \mu\}$, i.e. the set of all sample values for which $X_n(\omega)$ converges to μ, belongs to $\mathscr{A}$, and therefore has a *probability*. This probability is, in fact, equal to 1! $\bar{X}_n$ is said to *converge with probability 1* or *converge almost surely* to μ. This is the so-called *strong law of large numbers*:

$$P[\lim_{n \to \infty} \bar{X}_n = \mu] = 1. \tag{20}$$

These laws of large numbers play an important role in *mathematical statistics*. They state that the arithmetic mean of a large number of independent random variables with equal distribution lies close to the expected value of the individual random variables. The random variables

$$\sqrt{n} (\bar{X}_n - \mu)/\sigma = \left(\sum_{i=1}^{n} X_i - n\mu\right)\bigg/ \sqrt{n}\, \sigma \tag{21}$$

have the expected value 0 and variance 1. It can be shown that the distribution functions of the random variables (21) converge with increasing n to the *standardized normal distribution* (see Example (1.2.g)), i.e.

$$\lim_{n \to \infty} P\left[\frac{\sqrt{n} (\bar{X}_n - \mu)}{\sigma} \leqslant x\right] = \Phi(x) = \frac{1}{\sqrt{(2\pi)}} \int_{-\infty}^{x} e^{-z^2/2} \, dz. \tag{22}$$

This is a special case of the *central limit theorem*, one of the main results of probability theory.

For sufficiently large n we thus have

$$P[\sqrt{n}(\bar{X}_n - \mu)/\sigma \leqslant x] \approx \Phi(x), \tag{23}$$

where $\approx$ means 'is approximately equal to'. From this it follows that, if one replaces x in (23) by $\sqrt{n}(x - \mu)/\sigma$, then

$$P[\bar{X}_n \leqslant x] \approx \Phi(\sqrt{n}(x - \mu)/\sigma), \tag{24}$$

which means that, for sufficiently large n, $\bar{X}_n$ is approximately *normally distributed* with parameters μ and σ^2/n (see Examples (1.2.*g*) and (1.3.*h*)). It also follows that, if one replaces x by $(x - n\mu)\sigma\sqrt{n}$ in (23),

$$P\left[\sum_{i=1}^{n} X_i \leqslant x\right] \approx \Phi((x - n\mu)/\sigma\sqrt{n}). \tag{25}$$

The sum of the X_i is for sufficiently large n approximately *normally distributed* with the parameters $n\mu$ and $n\sigma^2$. In this sense $\bar{X}_n$ and ΣX_i are *asymptotically* normally distributed.

Example

(*d*) *Limit theorem of De-Moivre–Laplace.* Suppose the X_i, $i = 1, 2, \ldots$, are *independent Bernoulli variables* (see Example (1.2.*b*)), then the sum $X_1 + X_2 + \ldots + X_n$ is, on the one hand, according to Example (*b*), binomially distributed with parameters n and p. On the other hand, however, this sum is, according to (25), and for sufficiently large n, approximately normally distributed with parameters np and $np(1 - p)$. The distribution function of the binomial distribution, therefore, approaches asymptotically for large n the normal distribution with the corresponding parameters. This special case of the central limit theorem is due to De-Moivre–Laplace.   □

Notes to Chapter 1

This short introduction to general probability theory serves the needs of the later chapters. There are many excellent texts on probability theory which can be used for complementary reading. [1], [3, vol. 1] and [6] can be recommended as introductory accounts. [3] is a stimulating, applications-oriented, book on probability theory. [4], [5] and [7] contain systematic developments of the mathematical foundations of probability. [2] is a classical text on stochastic processes, but is still very readable.

2
Renewal theory

2.1. Renewal processes

Devices such as computers, aircraft, electronic appliances and others consist of parts which fail from time to time and have to be replaced or repaired. The time intervals between two failures are generally randomly distributed so that the failures themselves are randomly distributed on the time axis. The investigation of such failure and renewal phenomena is the reason for the introduction of so-called *renewal processes*. The study of such renewal processes is not only important in view of its application to the above mentioned physical phenomena. Renewal processes occur in connection with more complicated stochastic processes as well. Results from the theory of renewal processes, the so-called *renewal theory*, enable us to extract surprisingly extensive mathematical statements about stochastic processes. This double motivation, practical as well as theoretical, has been the reason for the development of renewal theory in this chapter.

Let $X_1, X_2, \ldots$ be a sequence of *independent* random variables which all have the *same* distribution function $F(x)$. Suppose that $F(0) = P[X_i \leq 0] = 0$, i.e. $F(x) = 0$ for all $x < 0$. The random variables X_i can thus only take on *positive* values. Suppose

$$S_0 = 0; \quad S_k = X_1 + X_2 + \ldots + X_k, \quad k \geq 1, \tag{1}$$

then the sequence of random variables $S_0, S_1, S_2, \ldots$ is called a *renewal process*. This process has the following physical interpretation: after time X_1 the first *renewal* takes place, and X_k is the time interval between the $(k-1)$th and kth renewal. S_k itself marks the time of the kth renewal. Since the time X_1 from 0 to the first renewal has the same distribution as the intervals between any two successive renewals, it can be assumed that the process starts immediately after a previous renewal. $F(x)$ is called a *renewal distribution* and the X_i are *renewal intervals*.

Since S_k is the sum of k independent random variables, the distribution function $F^{(k)}(x)$ of S_k is the k-fold convolution of F (see Section 1.4):

$$F^{(k)}(x) = F^{*k}(x). \tag{2}$$

One can associate with every renewal process $\{S_k, k \geqslant 0\}$ a *counting process* $N(t)$ by defining

$$N(t) = n, \quad \text{if } S_n \leqslant t, \ S_{n+1} > t, \ t \geqslant 0. \tag{3}$$

$N(t)$ obviously represents the number of renewals up to and including time t. Here $t = 0$ does not count as a renewal as it is sometimes taken to do in the literature; i.e. $N(0) = 0$. From (3) it follows that $N(t) \geqslant n$ if and only if $S_n \leqslant t$. Therefore

$$P[N(t) \geqslant n] = P[S_n \leqslant t] = F^{(n)}(t),$$

and thus (see (1.2.5))

$$P[N(t) = n] = F^{(n)}(t) - F^{(n+1)}(t). \tag{4}$$

Let us extend at this point the present definitions of renewal processes: for X_1 we allow an initial distribution $G(x)$ which may differ from $F(x)$. This generalization is of importance for applications where the renewal process does not start immediately after a renewal, but at a later time. The interval until the first renewal is thus distributed differently from the subsequent renewal intervals. One speaks in this case of a *delayed renewal process.* For the distribution $F^{(k)}$ of S_k we now obtain

$$\begin{aligned} F^{(1)}(x) &= G(x), \\ F^{(k)}(x) &= G(x) * F^{*(k-1)}(x) = F^{(k-1)}(x) * F(x) \quad \text{for } k \geqslant 2. \end{aligned} \tag{5}$$

Secondly we permit that $F(\infty) = \lim_{x \to \infty} F(x) < 1$, i.e. that F is an *incomplete* distribution (see Section 1.2). The defect $1 - F(\infty)$ corresponds to the probability of an infinitely long renewal interval. In this case it is possible that no further renewals take place.

Let $M(t)$ be the expected value of the number of renewals up to and including time t; then, since $\lim_{N \to \infty} NF^{(N)}(t) = 0$ (see the proof of Theorem 1 below),

$$M(t) = E[N(t)] = \sum_{k=1}^{\infty} kP[N(t) = k] = \sum_{k=1}^{\infty} k(F^{(k)}(t) - F^{(k+1)}(t))$$

$$= \sum_{k=1}^{\infty} F^{(k)}(t) = \sum_{k=1}^{\infty} G * F^{*(k-1)}(t), \quad \text{for } t \geqslant 0, \tag{6}$$

$$M(t) = 0 \quad \text{for } t < 0.$$

Here $F^{*0}(t) = 0$ for $t < 0$ and $F^{*0}(t) = 1$ for $t \geqslant 0$. The function $M(t)$ is called the *renewal function.*

Theorem 1
For every $t \geqslant 0$, $M(t) < \infty$.

Proof. From $F^{(k)} = F^{(k-1)} * F$, equation (5) and from the definition
of convolution (1.4.10) it follows that $F^{(k)}(t) \leqslant F^{(k-1)}(t)F(t)$. If this
inequality is iterated, we obtain $F^{(k)}(t) \leqslant G(t)(F(t))^{k-1}$. If $F(t) < 1$, (6) converges geometrically. If $F(t) = 1$ there exists an h such that $F^{*h}(t) < 1$.
Because of the assumption $F(0) = 0$ we conclude that the X_i must be positive
with probability 1. For sufficiently large h it is thus obvious that
$F^{*h}(t) = P[X_1 + X_2 + \ldots + X_h \leqslant t] < 1$. As above it follows that
$F^{*mh}(t) \leqslant (F^{*h}(t))^m$. From $F^{*n}(t) \leqslant F^{*(n-1)}(t)F(t) \leqslant F^{*(n-1)}(t)$ we conclude
that $F^{*n}(t)$ is monotonically non-increasing in n and, therefore, that
$F^{*mh}(t) \leqslant F^{*n}(t) \leqslant F^{*(m-1)h}(t)$ for $(m-1)h < n < mh$. Thus (6) converges
geometrically in every case and the theorem is proved. ∎

If we add the convolution relations $F^{(k+1)}(t) = F^{(k)}(t) * F(t)$ from $k = 1$ to
∞, we obtain $M(t) - G(t) = M(t) * F(t)$ or, expanded (see (1.4.12)),

$$M(t) = G(t) + \int_0^t M(t-x)\mathrm{d}F(x), \quad t > 0. \tag{7}$$

This is of the form

$$Z(t) = z(t) + \int_0^t Z(t-x)\mathrm{d}F(x), \quad t > 0, \tag{8}$$

and such equations are known as *renewal equations*. If $G = F$ then (7)
becomes

$$U(t) = \int_0^t [1 + U(t-x)]\mathrm{d}F(x), \quad t > 0, \tag{9}$$

by setting $M(t) = U(t)$. This equation has the following direct probabilistic
justification: up to t a renewal takes place only if $X_1 \leqslant t$. The first renewal
takes place at time x with probability $\mathrm{d}F(x)$. Given that at time $x \leqslant t$ the first
renewal takes place, then in the remaining interval of duration $t - x$ we
expect $U(t - x)$ further renewals. If we sum or integrate over x from 0 up to
t, then (9) follows. Such an argument is called a *renewal argument.* Renewal
arguments are based on the fact that after the first renewal, a second, new
renewal process starts which has *exactly* the same probabilistic properties as
the original process. If $f(t)$ is the expected value of a random quantity which
depends on the duration t of the original process, or the probability of an
event which in turn depends on the duration t of the process and if the first
renewal takes place at time $x \leqslant t$, then the corresponding expected value or
the probability for the second process with respect to the duration from x to
t is equal to $f(t - x)$.

Renewal argument can be applied to many situations in order to establish
renewal equations for certain expected values or probability functions. Since
we can prove powerful results about the solutions of renewal equations, in

particular with regard to their asymptotic behaviour as $t \to \infty$, these methods are very effective indeed.

The solution of the renewal equation (8) can be written explicitly in terms of the renewal function $U(t)$.

Theorem 2
If $z(t)$ is bounded then

$$Z(t) = z(t) + \int_0^t z(t-x)dU(x), \quad t > 0 \tag{10}$$

is the only solution of the renewal equation (8), which is bounded on every finite interval.

Proof. From (10) it follows for $s < \infty$ that

$$\sup_{0 \leqslant t \leqslant s} |Z(t)| \leqslant \sup_{0 \leqslant t \leqslant s} |z(t)| + \int_0^s \sup_{0 \leqslant t \leqslant s} |z(t)| dU(x)$$

$$= \sup_{0 \leqslant t \leqslant s} |z(t)|(1 + U(s)) < \infty,$$

(by Theorem 1). This proves the boundedness of $Z(t)$ on finite intervals. That $Z(t)$ is the solution of the renewal equation (8) is a consequence of the following relation (where $z(t) = 0$ and $Z(t) = 0$ for $t < 0$, so that (8) and (10) can be written as convolution relations (see (1.4.12)))

$$Z = z + z * U = z + z * \left(\sum_{k=1}^{\infty} F^{(k)} \right) = z + z * F + z * \sum_{k=2}^{\infty} F^{(k)}$$

$$= z + \left(z + z * \sum_{k=1}^{\infty} F^{(k)} \right) * F = z + (z + z * U) * F = z + Z * F.$$

In order to show this we used the various associative and distributive laws for convolutions (see (1.4.13), (1.4.14)).

If there exists a second solution $Z'(t)$ of the renewal equation (8), which is bounded on finite intervals, then the difference $V(t) = Z(t) - Z'(t)$ is again bounded on finite intervals and satisfies the equation $V = V * F$. By iteration of the equation it follows that $V = V * F^{*k}$ or

$$|V(t)| = \left| \int_0^t V(t-x)dF^{*k}(x) \right| \leqslant \sup_{0 \leqslant s \leqslant t} |V(s)| F^{*k}(t), \quad k = 1, 2, \ldots .$$

However, in the proof of Theorem 1 we saw that $F^{*k}(t) \to 0$ for $k \to \infty$. Therefore, $V(t) = 0$ for all t, and $Z(t)$ is the unique solution bounded on finite intervals. ∎

Examples

(a) *Poisson process.* If a non-delayed renewal process $\{S_k, k \geq 0\}$ has an exponential distribution $F(x) = 1 - e^{-\lambda x}$ for $x \geq 0, F(x) = 0$ for $x < 0, \lambda > 0$, as renewal distribution, then this process is called a Poisson process. Without difficulty we can prove by induction on k and with the help of the relation $F^{(k)} = F^{(k-1)} * F$ that

$$1 - F^{(k)}(x) = e^{-\lambda x} \sum_{j=0}^{k-1} \frac{(\lambda x)^j}{j!}.$$

For the counting process $N(t)$ we obtain, according to (4),

$$P[N(t) = n] = e^{-\lambda t} \frac{(\lambda t)^n}{n!}, \quad n = 0, 1, 2, \dots.$$

For every fixed t this is a *Poisson distribution* (see Example (1.2.d)). Furthermore (see Example (1.3.e))

$$U(t) = E[N(t)] = e^{-\lambda t} \sum_{n=1}^{\infty} n \frac{(\lambda t)^n}{n!} = e^{-\lambda t} \lambda t \sum_{n=0}^{\infty} \frac{(\lambda t)^n}{n!} = \lambda t,$$

and we can easily verify that the function λt is the solution of the renewal equation (9). $\qquad\Box$

(b) *Residual life, age, inspected life.* If a fixed time t is chosen, then the time interval from t to the next renewal, the time interval from the last renewal to t and the time interval from the last to the next renewal can be determined. The lengths of these intervals are given by $S_{N(t)+1} - t, t - S_{N(t)}$ and $S_{N(t)+1} - S_{N(t)}$ and are called (in that order) residual life, age, and inspected life. For the distribution of these random variables, renewal equations can be given. The reader is warned against the premature conclusion that the inspected life is distributed according to the renewal distribution F. By choosing an arbitrary inspection time one is likely to end up in long renewal intervals rather than short ones. Here we assume that $G = F$. For the following we recall that $N(t)$ is the number of renewals up to and including time t and that the integration from 0 to t is, according to Section 1.3, the same as integration over $(0, t]$.

Let $R_z(t) = P[S_{N(t)+1} - t > z]; R_z(t)$ is equal to the probability of no renewal in the interval $(t, t + z]$. This event occurs if either no renewal at all happens up to $t + z$, or if the first renewal takes place at $x \leq t$ and thereafter no renewal falls into the interval $(t, t + z]$. The probability of a first renewal at x is $dF(x)$. Suppose a renewal does take place at x; then the probability of no renewal between t and $t + z$ is equal to $R_z(t - x)$. The renewal argument thus leads to the renewal equation

$$R_z(t) = 1 - F(t + z) + \int_0^t R_z(t - x) dF(x).$$

Similarly, let $A_z(t) = P[t - S_{N(t)} \geqslant z]$, i.e. $A_z(t)$ is the probability of no renewal in the interval $(t - z, t)]$. Then $A_z(t) = R_z(t - z)$ for $z < t$.

Finally, let $L_z(t) = P[S_{N(t)+1} - S_{N(t)} > z]$. If $t \leqslant z$ the event will certainly take place if there is no renewal up to and including z; if $t > z$ then the event will certainly happen if there is no renewal up to and including t. In both cases there is the alternative of a first renewal before t such that the renewal argument leads, as above, to

$$L_z(t) = 1 - F(\max(z,t)) + \int_0^t L_z(t - x) \mathrm{d}F(x).$$

If the renewal distribution is an *exponential distribution* with parameter λ, i.e. if the renewal process is a Poisson process (see Example (a)), the renewal equation for the residual life reads

$$R_z(t) = e^{-\lambda(t+z)} + \int_0^t R_z(t - x)\lambda e^{-\lambda x} \mathrm{d}x.$$

It is easily verified that this equation has the solution

$$R_z(t) = e^{-\lambda z}.$$

Equally we have $A_z(t) = e^{-\lambda z}$. With the Poisson process the residual life as well as age have the same exponential distribution as the renewal distribution and this distribution is *independent* of time t. $\square$

(c) *Age-renewal policies.* It may be of advantage to replace machines, plants or parts, not just at their first failure but even before then as a preventive measure, when they have reached a particular age T. This raises the problem of finding the optimal renewal age T. If the appliance has failed before reaching age T, it must also be replaced, which we assume to cost the amount c_1. A planned renewal at age T is assumed to cost c_2, where $c_2 < c_1$ (otherwise a preventive renewal policy would not be advantageous).

For a planned renewal at age T the renewal intervals $X_1, X_2, \ldots$ form a renewal process with the renewal distribution $\bar{F}(x) = F(x)$ for $x < T$, and $\bar{F}(x) = 1$ for $x \geqslant T$, where $F(x)$ is the original life distribution. The choice of T must be based on an examination of costs. Let $\rho \geqslant 0$ be a discount rate so that the present value of the cost c which occurs at time t is equal to $ce^{-\rho t}$. Let $c(t)$ be the expected value of the present value of the total cost of all renewals up to t. If at time $x \leqslant t$ a first renewal takes place then the present value of its cost is

$$L(x) = \begin{cases} c_1 e^{-\rho x}, & \text{if } x < T, \\ c_2 e^{-\rho T}, & \text{if } x = T. \end{cases}$$

The expected value of the cost of all further renewals up to time t, discounted for time x, is clearly equal to $c(t - x)$, and, discounted for time 0, these costs will be equal to $e^{-\rho x}c(t-x)$. Thus the renewal argument leads to the renewal equation

$$c(t) = \int_0^t L(x)\,\mathrm{d}\bar{F}(x) + \int_0^t e^{-\rho x}c(t-x)\,\mathrm{d}\bar{F}(x)$$

for $c(t)$. This is a renewal equation of form (8), as can be seen by introducing the distribution function $G(x)$ defined by $\mathrm{d}G(x) = e^{-\rho x}\,\mathrm{d}\bar{F}(x)$. If $\rho > 0$ then $G(x)$ is evidently *incomplete*. If in addition we introduce the function

$$z(t) = \int_0^t L(x)\,\mathrm{d}\bar{F}(x) = \begin{cases} c_1 \int_0^t e^{-\rho x}\,\mathrm{d}F(x), & t < T, \\[2mm] c_1 \int_0^{T-0} e^{-\rho x}\,\mathrm{d}F(x) + c_2 e^{-\rho T}(1 - F(T-0)), \\[2mm] & t \geqslant T, \end{cases}$$

then the renewal equation can be written as

$$c(t) = z(t) + \int_0^t c(t-x)\,\mathrm{d}G(x). \qquad\qquad \square$$

If the renewal distribution F has a density function f, the renewal equation (9) reads

$$U(t) = \int_0^t [1 + U(t-x)]f(x)\,\mathrm{d}x. \qquad\qquad (11)$$

Suppose for a moment that $U(t)$ has a derivative $u(t)$; by differentiation of (11) we then obtain

$$u(t) = f(t) + \int_0^t u(t-x)f(x)\,\mathrm{d}x. \qquad\qquad (12)$$

This is a renewal equation of form (8) which has, according to Theorem 2, a unique solution which is bounded on every finite interval, provided that $f(x)$ is bounded. We can easily verify that

$$U(t) = \int_0^t u(x)\,\mathrm{d}x$$

is a solution of (11), if $u(x)$ is a solution of (12). Thus the renewal function $U(t)$ has indeed a *density function* $u(t)$ if the renewal distribution $F(x)$ has a bounded density function $f(x)$.

Renewal processes with *arithmetical* renewal distributions or *discrete* renewal processes are an important case. A distribution $F(x)$ is called *arithmetical*, if $F(x)$ is a step function with jumps, at the most, at $\lambda, 2\lambda, 3\lambda, \ldots$ where $\lambda > 0$. Since in this case the X_i are discrete random variables which can only take multiples of λ as values, renewals are also only possible at a multiple of λ. With a delayed renewal process it is assumed that the initial distribution G has jumps at multiples of the same value λ as the renewal distribution F. Without any loss in generality renewal processes with arithmetic renewal distributions can be understood as discrete renewal processes for which renewals can only occur at times $n = 1, 2, 3, \ldots$. In this case let $g_k = P[X_1 = k]$ for $k = 1, 2, \ldots, g_0 = 0$, $f_k = P[X_i = k]$ for $i \geqslant 2$ and $k = 1, 2, \ldots, f_0 = 0$ and $f_j^{(k)} = P[S_k = j]$ for $k \geqslant 1$ and $j \geqslant 1$. Note that $f_j^{(1)} = g_j$. Since the union of the disjoint events

$$[S_{k-1} = h, X_k = j - h], \quad h = 1, \ldots, j - 1$$

is equal to the event $[S_k = j]$ and the events $[S_{k-1} = h]$ and $[X_k = j - h]$ are mutually independent it follows that

$$f_j^{(k)} = f_1^{(k-1)}f_{j-1} + f_2^{(k-1)}f_{j-2} + \ldots + f_{j-1}^{(k-1)}f_1, \quad j = 1, 2, \ldots \text{ and } k \geqslant 2.$$
(13)

Equation (13) is, of course, nothing but the discrete version of convolution $F^{(k)} = F^{(k-1)} * F$ (see Example (1.4.a)).

The renewal function $M(t)$ of the discrete renewal process can only have *jumps* at the integer values $n = 1, 2, 3, \ldots$. If we define

$$m_k = M(k) - M(k-1), \quad k = 1, 2, \ldots, \quad m_0 = 0,$$

it follows that

$$M(t) = \sum_{k=1}^{[t]} m_k \quad \text{for } t \geqslant 1, \qquad M(t) = 0 \quad \text{for } t < 1,$$

where $[t]$ denotes the largest integer less than or equal to t. In the discrete case we can introduce the random variables $Z_k = 1, 2, \ldots$ which take value 1 if a renewal takes place at time k and 0 if there is no renewal. For the counting process we obtain

$$N(t) = \sum_{k=1}^{[t]} Z_k \quad \text{for } t \geqslant 1, \qquad N(t) = 0 \quad \text{for } t < 1$$

and (see (1.3.25))

$$M(t) = \sum_{k=1}^{[t]} E[Z_k],$$

so that $m_k = E[Z_k]$. If p_k is the probability of a renewal at time k, we obtain from $E[Z_k] = 1 \cdot p_k + 0 \cdot (1 - p_k) = p_k$ that $p_k = m_k$; m_k is, therefore, the probability of a renewal at time k. If a renewal takes place at k then this is either a *first* renewal or a renewal has taken place at $k - j, j = 1, \ldots, k - 1$, and the *next* renewal takes place after further time j. The probability for the first event is g_k, the probabilities for the second events are $m_{k-j}f_j, j = 1, \ldots, k - 1$. By summing it follows that

$$m_1 = g_1; \quad m_k = g_k + \sum_{j=1}^{k-1} m_{k-j} f_j, \quad k = 2, 3, \ldots. \tag{14}$$

This is the discrete version of the renewal equation (7). It can also be obtained by taking the differences of (7) for $t = k$ and $t = k - 1$ and noting that $dF(x) = f_k$ for $x = k, k = 1, 2, \ldots$, and $dF(x) = 0$ otherwise.

One main task of renewal theory is the examination of the asymptotic behaviour of the renewal function $M(t)$ for $t \to \infty$ and, more generally, the behaviour of the solutions of renewal equations for $t \to \infty$. The next two sections are devoted to these problems.

2.2 Renewal theorem for discrete renewal processes

Discrete renewal processes or renewal processes with arithmetic renewal distribution have in some cases to be treated differently from others. This is particularly true for the fundamental theorem of renewal theory, the renewal theorem. The discrete renewal processes form a genuine special case and demand special consideration. Therefore this section is devoted to discrete renewal processes.

The topics of this section give an opportunity to introduce the *method of generating functions* which has proved a useful aid for many problems in connection with discrete random variables and processes. If $a_k, k = 0, 1, 2, \ldots$, is a sequence of real numbers, then its *generating function* is defined by

$$A(z) = \sum_{k=0}^{\infty} a_k z^k. \tag{1}$$

If the a_k are bounded, then the series (1) converges for $|z| < 1$. Sometimes it is advantageous to consider z as a complex variable.

Examples

(*a*) *Expected values.* Let X be a discrete random variable which can take values $0, 1, 2, \ldots$ and let

$$P[X = j] = p_j, \quad P[X > j] = q_j.$$

The two sequences $\{p_j, j = 0, 1, \ldots\}$ and $\{q_j, j = 0, 1, \ldots\}$ have the generating functions

$$P(z) = \sum_{j=0}^{\infty} p_j z^j, \quad Q(z) = \sum_{j=0}^{\infty} q_j z^j$$

and since $P(1) = 1$, the sequence $P(z)$ converges for $-1 \leqslant z \leqslant 1$, while $Q(z)$ converges at least for $-1 < z < 1$. For $-1 < z < 1$,

$$(1 - z)Q(z) = q_0 + \sum_{j=1}^{\infty} (q_j - q_{j-1})z^j.$$

According to the definition of q_j, $q_0 = 1 - p_0$ and $q_j - q_{j-1} = -p_j$. By substituting it follows that

$$Q(z) = (1 - P(z))/(1 - z), \quad -1 < z < 1.$$

Furthermore, by differentiating the series $P(z)$ with respect to z, we obtain

$$P'(z) = \sum_{j=1}^{\infty} j p_j z^{j-1},$$

where $P'(z)$ converges for the moment at least for $-1 < z < 1$. We note, however, that $P'(z)$ also converges for $z = 1$, if $E[X] < \infty$. If we let the variable $z \to 1$ in the above formula for $Q(z)$ we obtain $P'(1) = Q(1)$ so that

$$E[X] = \sum_{j=0}^{\infty} q_j.$$

Compare this result also with formula (1.3.38). $\qquad \square$

(*b*) *Convolutions.* If a_k, b_k, c_k are three sequences which are connected by the relations .

$$c_k = \sum_{j=0}^{k} a_{k-j} b_j, \quad k = 0, 1, 2, \ldots,$$

then the sequence c_k is the *convolution* of the sequences a_k and b_k. A coefficient comparison between the generating functions $C(z)$ and $A(z)B(z)$ shows that $C(z) = A(z)B(z)$. The generating function of the convolution of two sequences is equal to the product of the generating functions of the two sequences. If X_1, X_2 are two independent random variables with the possible values $0, 1, 2, \ldots$ and if $P_1(z), P_2(z)$ are the generating functions of their probabilities as in Example (*a*), then $P(z) = P_1(z)P_2(z)$ is the generating function of the probabilities of the sum $X = X_1 + X_2$, since the probabilities of X are obtained by the convolution of the probabilities of X_1 and X_2 (see Example (1.4.*a*)). $\qquad \square$

In connection with a discrete renewal process the following generating functions can be introduced:

$$G(z) = \sum_{k=0}^{\infty} g_k z^k, \qquad F(z) = \sum_{k=0}^{\infty} f_k z^k,$$

$$M(z) = \sum_{k=0}^{\infty} m_k z^k, \qquad F^{(r)}(z) = \sum_{k=0}^{\infty} f_k^{(r)} z^k,$$

where we recall that $g_0 = f_0 = m_0 = 0$. Since the $g_k, f_k, f_k^{(r)}$ determine distributions which are possibly incomplete, we have $g = G(1) \leqslant 1$, $f = F(1) \leqslant 1$, $F^{(r)}(1) \leqslant 1$. Therefore $G(z), F(z), F^{(r)}(z)$ converge for $-1 \leqslant z \leqslant 1$. $1 - f = P[X_i = \infty]$ is the probability that the renewal interval is infinitely long and that no more renewals take place. The m_k are probabilities and therefore $M(z)$ converges at least for $-1 < z < 1$. According to Example (b) and $(2.1.5)$, $F^{(r)}(z) = G(z)(F(z))^{r-1}$. From this we conclude in particular for $z = 1$ that

$$\sum_{k=0}^{\infty} f_k^{(r)} = g f^{r-1}.$$

On the other hand, however, $f_0^{(r)} + f_1^{(r)} + f_2^{(r)} + \ldots = P[S_r < \infty]$. If $g = f = 1$ then S_r is finite with probability 1 for all $r = 1, 2, \ldots$, i.e. with probability 1 there are *infinitely many renewals*. If on the other hand $f < 1$ then S_r is infinite with probability $1 - g f^{r-1} > 0$ and this probability tends to 1 for $r \to \infty$. If $g = 1$ then

$$P[S_r < \infty, S_{r+1} = \infty] = P[S_r < \infty, X_{r+1} = \infty] = f^{r-1}(1 - f)$$

is the probability that *exactly* r renewals occur. Thus, the probability that only *finitely* many renewals occur is equal to

$$\sum_{r=1}^{\infty} f^{r-1}(1 - f) = 1,$$

if $f < 1$. For this reason we call a renewal process with $f = 1$ *recurrent* and one with $f < 1$ *transient*.

If we multiply the renewal equations $(2.1.14)$ for k by z^k and sum them over $k = 1, 2, \ldots$ the result about convolutions, contained in Example (b), yields

$$M(z) = G(z) + M(z)F(z)$$

or

$$M(z) = G(z)/(1 - F(z)). \tag{2}$$

With the help of relation (2) the following criterion for recurrence or transience of a renewal process can be proved.

Theorem 1

(a) A renewal process is recurrent if and only if

$$m = \sum_{k=1}^{\infty} m_k = \infty.$$

(b) A renewal process is transient if and only if

$$m = \sum_{k=1}^{\infty} m_k < \infty.$$

In this case, if $g = 1$,

$$f = (m - 1)/m. \tag{3}$$

Proof. Since $m_k \geqslant 0$ and $M(z)$ is consequently monotonically increasing for $z \to 1$, we have for every N

$$\sum_{k=1}^{N} m_k \leqslant \lim_{z \to 1} M(z) \leqslant \sum_{k=1}^{\infty} m_k \tag{4}$$

and therefore $m = \lim_{z \to 1} M(z)$.

If $f < 1$, then it follows from (2) that $\lim_{z \to 1} M(z) = g/(1-f) < \infty$ and thus $m = g/(1-f) < \infty$. This also implies (3). On the other hand, if $f = 1$, it follows from (2) that $\lim_{z \to 1} M(z) = \infty$ and thus $m = \infty$. ∎

A discrete renewal process is called *periodic* if there is an integer $d > 1$ such that $f_k \neq 0$ only for $k = d, 2d, 3d, \ldots$, i.e. if renewals can only take place in intervals which are multiples of d. The largest integer d with this property is called the *period* of the renewal process. A non-periodic renewal process is called *aperiodic*.

It follows from Theorem 1 that for a transient renewal process $m_k \to 0$ for $k \to \infty$. The following *renewal theorem for discrete renewal processes* shows that the m_k have a limit also in the case of recurrence.

Theorem 2

For every aperiodic recurrent renewal process we have

$$\lim_{k \to \infty} m_k = g/\mu, \tag{5}$$

where

$$\mu = \sum_{j=1}^{\infty} j f_j \tag{6}$$

is the expected value of the renewal distribution. If $\mu = \infty$, we put $g/\mu = 0$.

Proof. We have shown at the end of Section 2.1 that the m_k are probabilities. Consequently $0 \leqslant m_k \leqslant 1$. This can be proved analytically from the renewal equation (2.1.14) by induction on k. Thus there exists $\bar{m} = \lim \sup_{k \to \infty} m_k$ and $\underline{m} = \lim \inf_{k \to \infty} m_k$ and there exist subsequences m_{k_ν} which tend to $\bar{m}$ and others which converge to $\underline{m}$. For the remainder of the proof we require the following lemma.

Lemma

Under the assumptions of the theorem we have:
(a) If $m_{k_\nu} \to \bar{m}$ $(m_{k_\nu} \to \underline{m})$ for $k_\nu \to \infty$, then $m_{k_\nu-j} \to \bar{m}$ $(m_{k_\nu-j} \to \underline{m})$ where j is fixed and chosen such that $f_j > 0$.
(b) Among the indices j with $f_j > 0$ there exists a finite set of relatively prime indices $j_1, j_2, \ldots, j_m$, (i.e. $j_1, j_2, \ldots, j_m$ have highest common factor 1).
(c) If $m_{k_\nu} \to \bar{m}$ $(m_{k_\nu} \to \underline{m})$ for $k_\nu \to \infty$, then $m_{k_\nu-k} \to \bar{m}$ $(m_{k_\nu-k} \to \underline{m})$

for all $k > j_1 j_2 \ldots j_m$, if $j_1, j_2, \ldots, j_m$ are relatively prime and $f_{j_i} > 0$ for $i = 1, 2, \ldots, m$.

The proof of the lemma will be given after the end of the proof of the theorem which we now continue. Let

$$r_n = f_{n+1} + f_{n+2} + \ldots, \quad n = 0, 1, 2, \ldots.$$

According to Example (a), $\mu = r_0 + r_1 + r_2 + \ldots$. We have $r_0 = 1$ and $f_k = r_{k-1} - r_k, k = 1, 2, \ldots$. If we insert this in (2.1.14) it follows that

$$r_0 m_k + r_1 m_{k-1} + \ldots + r_{k-1} m_1 = r_0 m_{k-1} + r_1 m_{k-2} + \ldots$$
$$+ r_{k-2} m_1 + g_k. \tag{7}$$

If the left hand side of (7) is called A_k, then (7) reads $A_k = A_{k-1} + g_k$ and $A_0 = r_0 m_0 = 0$. From this we obtain

$$A_k = \sum_{j=0}^{k} g_k. \tag{8}$$

Let $\{j_1, j_2, \ldots, j_m\}$ be a set of relatively prime indices with $f_{j_i} > 0$ (which exists according to the lemma) and let $k > j_1 j_2 \ldots j_m$ but otherwise arbitrary. Furthermore, let $m_{k_\nu} \to \bar{m}$. From

$$r_0 m_{k_\nu - k} + r_1 m_{k_\nu - k - 1} + \ldots + r_N m_{k_\nu - k - N} \leq A_{k_\nu - k} = \sum_{j=0}^{k_\nu - k} g_j \leq g$$

and from the lemma it follows for $k_\nu \to \infty$ and for $N < k_\nu - k - 1$ that

$$\bar{m}(r_0 + r_1 + \ldots + r_N) \leq g.$$

Since N is arbitrary here and $(r_0 + r_1 + \ldots + r_N) \to \mu$ for $N \to \infty$, we conclude that $\bar{m}\mu \leq g$, or $\bar{m} \leq g/\mu$. If $\mu = \infty$ the assertion follows, since $0 \leq \underline{m} \leq \bar{m} \leq 0$ and thus $m_k \to 0$ for $k \to \infty$.

Otherwise if $\mu < \infty$ and $m_{k_\nu} \to \underline{m}$, we obtain

$$r_0 m_{k_\nu - k} + r_1 m_{k_\nu - k - 1} + \ldots + r_N m_{k_\nu - k - N} + \epsilon \geq \sum_{j=0}^{k_\nu - k} g_j,$$

where N is chosen large enough so that $(r_{N+1} + r_{N+2} + \ldots) < \epsilon$. For $k_\nu \to \infty$ it follows from the lemma as before that

$$\underline{m}(r_0 + r_1 + \ldots + r_N) + \epsilon \geq \sum_{j=0}^{\infty} g_j = g,$$

from which we conclude that $\underline{m}\mu \geq g$, or $\underline{m} \geq g/\mu$. Since $g/\mu \leq \underline{m} \leq \bar{m} \leq g/\mu$ the statement of the theorem also follows for $\mu < \infty$. ∎

Proof of the lemma. (a) If (a) were invalid there would exist arbitrarily large k such that $m_k > \bar{m} - \epsilon$ and $m_{k-j} \leq m < \bar{m}$. If one chooses

$N > j$ sufficiently large such that $f_{N+1} + f_{N+2} + \ldots < \epsilon$, then it follows from (2.1.14) for $k > N$, since $m_n \leqslant 1$, that

$$m_k \leqslant g_k + f_1 m_{k-1} + \ldots + f_N m_{k-N} + \epsilon.$$

k can be chosen sufficiently large so that $m_{k-h} < \bar{m} + \epsilon$ for $h = 1, \ldots, N$ and $g_k < \epsilon$. Then

$$m_k \leqslant (f_1 + \ldots + f_{j-1} + f_{j+1} + \ldots + f_N)(\bar{m} + \epsilon) + f_j m + 2\epsilon$$
$$\leqslant (1 - f_j)(\bar{m} + \epsilon) + f_j m + 2\epsilon$$
$$< \bar{m} - f_j(\bar{m} - m) + 3\epsilon.$$

If we choose ϵ such that $f_j(\bar{m} - m) > 4\epsilon$ then $m_k < \bar{m} - \epsilon$ which contradicts our assumption that $m_k > \bar{m} - \epsilon$. Therefore we cannot have $m < \bar{m}$ and (a) must hold for $\bar{m}$. The second part of (a) concerning m is proved in a similar manner.

(b) must hold since the renewal process would otherwise have to be periodic, contrary to our assumption. If $f_1 > 0$ the set only consists of $j_1 = 1$.

(c) If $m_{k_\nu} \to \bar{m}$ and if $f_j > 0$, then it follows by repeated application of (a) that $m_{k_\nu - 2j} \to \bar{m}$, $m_{k_\nu - 3j} \to \bar{m}$, etc., hence, that $m_{k_\nu - h_1 j_1} \to \bar{m}$, $m_{k_\nu - h_2 j_2} \to \bar{m}$, $\ldots, m_{k_\nu - h_m j_m} \to \bar{m}$, where $f_{j_i} > 0$, and finally, again through repeated application of (a), that $m_{k_\nu - h_1 j_1 - h_2 j_2 - \ldots - h_m j_m} \to \bar{m}$. Every integer $k > j_1 j_2 \ldots j_m$ can, however, be represented in the form $k = h_1 j_1 + h_2 j_2 + \ldots + h_m j_m$, provided that $j_1, j_2, \ldots, j_m$ are relatively prime. Thus we have also proved (c). ∎

The reader should be aware of the fact that the proof of Theorem 2 only uses the fact that the m_k are solutions of the renewal equations (2.1.14). *Thus Theorem 2 holds generally for solutions of equations of the form (2.1.14).*

As a corollary of Theorem 2 we obtain a simple limit theorem for periodic renewal processes.

Corollary

For a recurrent non-delayed renewal process of period $d > 1$, we have

$$\lim_{k \to \infty} m_{dk} = d/\mu,$$

where μ is defined by (6).

Proof. $S_k' = S_k/d$ is a new aperiodic renewal process with renewal intervals $X_i' = X_i/d$ and $\mu' = E[X_i'] = \mu/d$ and $m_k' = m_{dk}$. The assertion thus follows from Theorem 2 applied to the process S_k'. ∎

Examples

(c) *Residual life, age.* Let us consider a discrete, aperiodic and recurrent renewal process with renewal distribution $\{f_k, k = 1, 2, \ldots\}$. If j is a fixed time, $j = 0, 1, 2, \ldots$, then one can consider the probabilities $r_k^{(j)}$ that the next renewal will take place at time $j + k$. The time interval from an arbitrary time j to the next renewal is called residual life (see Example (2.1.b)) and $\{r_k^{(j)}, k = 1, 2, \ldots\}$ is the distribution of the residual life at time j. The next renewal occurs at time $j + k$, if this is the very first renewal (probability f_{j+k}), or if a first renewal has occurred at time $i \leqslant j$ (a possible renewal at j occurs *before* the inspection) and the next renewal after j occurs at time $j + k$. The probability for this is $f_i r_k^{(j-i)}$. By adding the probabilities of these disjoint events it follows that

$$r_k^{(j)} = f_{j+k} + \sum_{i=1}^{j} f_i r_k^{(j-i)}, \quad j = 0, 1, 2, \ldots .$$

For every fixed $k = 1, 2, \ldots$ this is a renewal equation of form (2.1.14) where the $r_k^{(j)}$ correspond to the m_{j+1} and the f_{j+k} correspond to the g_{j+1}. Therefore the renewal theorem, Theorem 2, can be applied:

$$\lim_{j \to \infty} r_k^{(j)} = \frac{1}{\mu} \sum_{j=0}^{\infty} f_{j+k} = \frac{1}{\mu}\left(1 - \sum_{j=1}^{k-1} f_j\right), \quad k = 1, 2, \ldots .$$

Here μ is the expected value of the renewal distribution.

In addition let $a_k^{(j)}$ be the probability that the last renewal before time j occurred at time $j - k$. Then k is called the age at time j (see also Example (2.1.b)). Furthermore, let $s_k^{(j)}$ be the probability that at time j a renewal takes place and that the next renewal will occur at the earliest at time $j + k$. It is obvious that $a_k^{(j)} = s_k^{(j-k)}$, where, in contrast to Example (2.1.b) and the above convention for the residual life, the inspection at time j is assumed to be *before* a possible renewal at j. However, $s_k^{(j)}$ satisfies the renewal equation

$$s_k^{(j)} = f_j\left(1 - \sum_{i=1}^{k-1} f_i\right) + \sum_{i=1}^{j-1} f_i s_k^{(j-i)}, \quad j = 1, 2, \ldots ,$$

where the first term corresponds to the case in which the first renewal takes place at j, while the second term considers the cases of a first renewal at $i < j$. From the renewal theorem (Theorem 2), it thus follows for the limit distribution of age that

$$\lim_{j \to \infty} s_k^{(j)} = \lim_{j \to \infty} a_k^{(j)}$$

$$= \frac{1}{\mu} \sum_{j=1}^{\infty} f_j\left(1 - \sum_{i=1}^{k-1} f_i\right) = \frac{1}{\mu}\left(1 - \sum_{i=1}^{k-1} f_i\right), \quad k = 1, 2, \ldots ,$$

i.e. we obtain the same limit distribution as for the residual life. □

2.3. Renewal theorem for continuous renewal processes

In this section we exclude arithmetic renewal distributions F. Let $F(\infty) = \lim_{x \to \infty} F(x)$. The renewal distribution is incomplete with defect $1 - F(\infty) > 0$ if $F(\infty) < 1$. In this case the probability that altogether precisely r renewals will take place before the process terminates is equal to $F^r(\infty)(1 - F(\infty))$, where $r = 0, 1, 2, \ldots$. We assume a non-delayed renewal process, i.e. $G = F$. It is easy to extend this consideration to the general case of a delayed renewal process and it is left to the reader to do so. The probability that only *finitely* many renewals occur is thus

$$\sum_{r=0}^{\infty} F^r(\infty)(1 - F(\infty)) = 1.$$

For $F(\infty) < 1$ the process terminates with probability 1 after finitely many renewals. Analogously to the discrete case the process is called *transient* in this case. Transient processes are easily dealt with and shall therefore be considered first.

The expected value of the number of renewals up to time t is given by the renewal function

$$U(t) = \sum_{k=1}^{\infty} F^{(k)}(t).$$

$F^{(k)}(\infty) = \lim_{t \to \infty} F^{(k)}(t)$ is the probability that *at least* k renewals take place and thus $F^{(k)}(\infty) = F^k(\infty)$. It follows that $U(t)$ converges for $t \to \infty$ to the finite value

$$U(\infty) = \sum_{k=1}^{\infty} F^k(\infty) = \frac{F(\infty)}{1 - F(\infty)} < \infty.$$

Theorem 2.1.2 also applies to transient renewal processes. If $z(t)$ is a bounded function for which $z(\infty) = \lim_{t \to \infty} z(t)$ exists, then the solution $Z(t)$ of the renewal equation (2.1.10) satisfies

$$\lim_{t \to \infty} Z(t) = \frac{z(\infty)}{1 - F(\infty)}. \tag{1}$$

We prove this by subdividing the integral in (2.1.10) into one from 0 to t' and one from t' to t. We choose t' sufficiently large such that $U(\infty) - U(t') < \epsilon$ and t sufficiently large such that $|z(\infty) - z(t - x)| < \epsilon$ for $0 \leqslant x \leqslant t'$. Thus the first integral only differs by ϵ from $z(\infty)U(\infty)$ and the second by ϵ from 0. The details are left to the reader.

Example

(*a*) *Age-renewal policies.* We continue from Example (2.1.*c*). In the case of a *positive* discount factor $\rho > 0$ the distribution $G(x)$ which appears

in the renewal equation for the expected value of the discounted cost $c(t)$, is
incomplete. Furthermore, the limit of $z(t)$ for $t \to \infty$ exists:

$$\lim_{t \to \infty} z(t) = z(T) = c_1 \int_0^{T-0} e^{-\rho x} dF(x) + c_2 e^{-\rho T}(1 - F(T-0)).$$

Thus (1) can be applied and we obtain for the expected value of the discounted cost of an age-renewal policy T over an infinite period

$$c(\infty) = \lim_{t \to \infty} c(t) = \frac{z(T)}{1 - G(T)},$$

since

$$G(\infty) = \lim_{t \to \infty} G(t) = \lim_{t \to \infty} \int_0^t e^{-\rho x} d\bar{F}(x)$$

$$= \int_0^{T-0} e^{-\rho x} dF(x) + e^{-\rho T}(1 - F(T-0)) = G(T).$$

For a given life distribution $F(x)$ the optimal age T for an age-renewal policy
T which minimizes the expected value $c(\infty)$ of the discounted cost can easily
be determined. It is not possible that the optimum occurs at $T = \infty$ which
means that no preventive age-renewal policy is worth while.

We may ask which constant average cost rate $m(\rho)$ (cost per unit time)
might lead to the same discounted cost $c(\infty)$. m must obviously satisfy the
equation

$$c(\infty) = \int_0^\infty m(\rho) e^{-\rho t} dt = \frac{m(\rho)}{\rho},$$

i.e. $m(\rho) = \rho z(T)/(1 - G(T))$. If we differentiate the numerator and denominator of m with respect to ρ and if we let $\rho \to 0$, then, according to the rule
of l'Hôpital

$$\lim_{\rho \to 0} m(\rho) = \frac{c_1 F(T-0) + c_2(1 - F(T-0))}{\int_0^T x d\bar{F}(x)}.$$

This limit obviously corresponds to the expected value of the average cost
per unit time for the age-renewal policy T *without* discounting. This will be
confirmed later in Example (c) in a different way.   □

In the following we assume $F(\infty) = 1$. As in the discrete case (Section 2.2)
we can show that with probability 1 *infinitely* many renewals occur. These
are therefore *recurrent* renewal processes. The renewal function $U(t)$ for
recurrent processes has unlimited growth. In the following Theorem 2 we
shall describe the asymptotic behaviour of the renewal function for $t \to \infty$.
Before that we shall show that $U(t + h) - U(t)$ is bounded as a function of t
for every value of $h > 0$ (see also Theorem 2.1.1).

Theorem 1
For every $t > 0$

$$U(t + h) - U(t) \leqslant 1 + U(h). \tag{2}$$

Proof. Let $N(t, h) = N(t + h) - N(t)$ be the number of renewals in the interval $(t, t + h]$. If $S_{N(t)+n+1} > t + h$, then certainly $N(t, h) \leqslant n$. On the other hand $S_{N(t)+n+1} > t + h$ whenever $X_{N(t)+2} + \ldots + X_{N(t)+n+1} > h$. The distribution of $X_{N(t)+2} + \ldots + X_{N(t)+n+1}$ is, however, the same as that of S_n. Therefore $P[N(t, h) \leqslant n] \geqslant P[S_n > h]$ (see (1.1.8)), or

$$P[N(t, h) > n] \leqslant P[S_n \leqslant h].$$

Thus (see Example (2.2.*a*) and (2.1.6))

$$U(t + h) - U(t) = \sum_{n=0}^{\infty} P[N(t, h) > n] \leqslant \sum_{n=0}^{\infty} P[S_n \leqslant h] = 1 + U(h). \qquad \blacksquare$$

Now we can formulate the *renewal theorem* for non-discrete processes. This is the analogue of Theorem 2.2.2 for discrete renewal processes.

Theorem 2
If the renewal distribution F is *not* arithmetic then, for $h > 0$,

$$U(t + h) - U(t) \to h/\mu \quad \text{for } t \to \infty, \tag{3}$$

where μ is the expected value

$$\mu = \int_0^{\infty} x \, dF(x) \tag{4}$$

of the *renewal distribution* and h/μ is set equal to 0 if $\mu = \infty$.

Proof. This is a purely analytic proof and does not give any insight into the probabilistic nature of the theorem. Therefore the proof is only presented very briefly. It is based on two analytic results which are stated in the appropriate place as lemmas without proof.

Let $g(t)$ be a continuous bounded function which vanishes outside $[0, h]$ and let, for $t > 0$,

$$\phi(t) = g(t) + \int_0^t g(t - s) \, dU(s)$$

$$= g(t) + \int_{t-h}^t g(t - s) \, dU(s).$$

By using (2) it follows that

$$|\phi(t)| \leqslant \sup_{0 \leqslant x \leqslant h} |g(x)|(2 + U(h))$$

and for $\delta > 0$ we obtain

$$|\phi(t + \delta) - \phi(t)| \leqslant \sup_{-\delta \leqslant x \leqslant h} |g(x + \delta) - g(x)|(2 + U(h + \delta)).$$

The function $\phi(t)$ is thus bounded and uniformly continuous. According to Theorem 2.1.2 $\phi(t)$ is, furthermore, the solution of the renewal equation

$$\phi(t) = g(t) + \int_0^t \phi(t - s) \, dF(s). \tag{5}$$

If we integrate this equation from 0 to $t > h$ then, using the formula for partial integration (1.3.36), we obtain

$$\int_0^h g(s) \, ds = \int_0^t \phi(s) \, ds - \int_0^t ds \int_0^s \phi(s - x) \, dF(x)$$

$$= \int_0^t \phi(t - s) \, ds - \int_0^t dF(x) \int_0^{t-x} \phi(s) \, ds$$

$$= \int_0^t \phi(t - s) \, ds - \left[F(x) \int_0^{t-x} \phi(s) \, ds \right]_{x=0}^t - \int_0^t F(x) \phi(t - x) \, dx$$

$$= \int_0^t \phi(t - s)(1 - F(s)) \, ds. \tag{6}$$

Suppose $\bar{m} = \lim \sup_{t \to \infty} \phi(t)$ and t_n is a subsequence such that $\phi(t_n) \to \bar{m}$. We define a sequence of functions by $z_n(x) = \phi(t_n + x)$ for $-t_n < x < \infty$, $z_n(x) = 0$ for $x \leqslant -t_n$. It follows from the uniform continuity of $\phi(t)$ that, for every $\epsilon > 0$, there exists a $\delta > 0$ such that for all x', x'' with $|x' - x''| < \delta$ we have

$$|\phi(t_n + x') - \phi(t_n + x'')| = |z_n(x') - z_n(x'')| < \epsilon, \quad \text{for all } n.$$

A family of functions $\{z_n(x)\}$ with this property is called *equicontinuous*. We have

Lemma 1

   If $\{z_n(x)\}$ is a sequence of equicontinuous functions with $|z_n(x)| \leqslant C$, there exists a subsequence $\{z_{n_k}(x)\}$ which converges to a continuous function $z(x)$. This convergence is uniform on bounded intervals (the proof is omitted; see Notes to Chapter 2).

From (5) we note that the limit function $z(x)$, obtained by applying the lemma, satisfies the equation

$$z(x) = \int_0^\infty z(x-s)\,dF(s), \quad -\infty < x < \infty. \tag{7}$$

Furthermore, we have

Lemma 2

Every bounded and continuous solution of (7) is a constant if F is not arithmetic (the proof is omitted; see Notes to Chapter 2).

Thus $z(x) = z(0) = \bar{m}$. If t_n is a subsequence for which $\phi(t_n + x) \to \bar{m}$ for all x, then it follows from (6) that

$$\lim_{n \to \infty} \int_0^{t_n} \phi(t_n - s)(1 - F(s))\,ds = \bar{m} \int_0^\infty (1 - F(s))\,ds$$

$$= \bar{m}\mu = \int_0^h g(s)\,ds$$

(see (1.3.38)) and thus

$$\bar{m} = \frac{1}{\mu} \int_0^h g(s)\,ds.$$

If $\underline{m} = \liminf_{t \to \infty} \phi(t)$, we show similarly that

$$\underline{m} = \frac{1}{\mu} \int_0^h g(s)\,ds.$$

It follows that $\bar{m} = \underline{m} = \lim_{t \to \infty} \phi(t)$ or

$$\lim_{t \to \infty} \left\{ g(t) + \int_{t-h}^t g(t-s)\,dU(s) \right\} = \lim_{t \to \infty} \int_{t-h}^t g(t-s)\,dU(s)$$

$$= \frac{1}{\mu} \int_0^h g(s)\,ds \tag{8}$$

for every continuous function $g(s)$ which vanishes outside $[0, h]$. The right hand side of (8) is equal to zero for $\mu = \infty$.

Let $0 < h_1 < h_2 < h$, $\delta > 0$, but sufficiently small such that $h_1 - \delta > 0$, $h_2 + \delta < h$, and $g(s)$ is a continuous function with $g(s) = 1$ in the interval $[h_1, h_2]$, $g(s) = 0$ outside the interval $[h_1 - \delta, h_2 + \delta]$ and $0 \leqslant g(s) \leqslant 1$ otherwise. Then the right hand side of (8) is less than $(h_2 - h_1 + 2\delta)/\mu$. The integral on the left hand side is, on the other hand, greater than $U(t - h_1) - U(t - h_2)$. If $\epsilon > 0$ and t is sufficiently large, we have

$$U(t - h_1) - U(t - h_2) \leqslant \int_{t-h}^t g(t-s)\,dU(s)$$

$$< \frac{1}{\mu} \int_0^h g(s)\,ds + \epsilon < \frac{1}{\mu}(h_2 - h_1 + 2\delta) + \epsilon. \tag{9}$$

Suppose next that $g(s)$ is continuous and $g(s) = 1$ on $[h_1 + \delta, h_2 - \delta]$, $g(s) = 0$ outside $[h_1, h_2]$, $0 \leqslant g(s) \leqslant 1$ otherwise. The right hand side of (8) is greater than $(h_2 - h_1 - 2\delta)/\mu$ and the integral on the left hand side is less than $U(t - h_1) - U(t - h_2)$. For sufficiently large t we thus have

$$U(t - h_1) - U(t - h_2) \geqslant \int_{t-h}^{t} g(t - s) dU(s)$$

$$> \frac{1}{\mu} \int_0^h g(s) ds - \epsilon > \frac{1}{\mu} (h_2 - h_1 - 2\delta) - \epsilon. \tag{10}$$

Since δ, ϵ are arbitrary we obtain from (9) and (10)

$$\lim_{t \to \infty} \{U(t - h_1) - U(t - h_2)\} = (h_2 - h_1)/\mu$$

and this proves the theorem. ∎

In order to compare discrete and non-discrete renewal processes let us note that in the discrete case $m_n = U(n) - U(n - 1)$ and that the renewal theorem (Theorem 2.2.2), for aperiodic discrete processes implies that $m_n = U(n) - U(n - 1) \to 1/\mu$ as $n \to \infty$. This implies furthermore that $m_{n+h} + \ldots + m_{n+1} = U(n + h) - U(n) \to h/\mu$ as $n \to \infty$. For discrete aperiodic renewal processes (3) holds if t and h are restricted to integers but, in contrast, not for arbitrary values of h.

As a corollary to the renewal theorem we can obtain a result which is often called the *elementary renewal theorem*. The theorem is called elementary because it can also be directly proved without the use of the renewal theorem (Theorem 2). In our proof, however, we shall make use of the renewal theorem. The corollary holds also for *discrete* renewal processes.

Corollary
If $\mu < \infty$ we have

$$U(t)/t \to 1/\mu \quad \text{as } t \to \infty. \tag{11}$$

Proof. Let $a_n = U(n) - U(n - 1)$. In the non-arithmetic case it follows from Theorem 2 that $a_n \to 1/\mu$ as $n \to \infty$. The arithmetic mean of the a_n is equal to $U(n)/n$ and has also to converge to $1/\mu$. This assertion follows from

$$\frac{[t]}{t} \frac{U([t])}{[t]} \leqslant \frac{U(t)}{t} \leqslant \frac{[t] + 1}{t} \frac{U([t] + 1)}{[t] + 1},$$

where $[t]$ denotes the largest integer less than t.

For discrete renewal processes, $a_n = m_n$. The assertion in this case follows from Theorem 2.2.2 (aperiodic case), or from the corollary to Theorem 2.2.2 (periodic case). ∎

With the help of the renewal theorem we are able to make statements about the asymptotic behaviour of solutions $Z(t)$ (2.1.10) of the renewal equation (2.1.8) for a certain class of functions $z(t)$. If $z(x) = 1$ for $0 \leqslant a \leqslant x < b < \infty$ and $z(x) = 0$ for all other x, it follows from (2.1.10) and Theorem 2 for $t > b$ and $t \to \infty$ that

$$Z(t) = U(t-a) - U(t-b) \to (b-a)/\mu. \tag{12}$$

If more generally $[a_k, b_k)$, $k = 1, 2, \ldots$, are s disjoint intervals of length $h_k = b_k - a_k$, and if $z_k(x) = 1$ for $a_k \leqslant x < b_k$ and $z(x) = 0$ otherwise, then $z(x) = c_1 z_1(x) + c_2 z_2(x) + \ldots + c_s z_s(x)$ is a *step function*. Let $Z_k(t)$ be the solution for the function $z_k(t)$. The solution $Z(t)$ for $z(t)$ then satisfies the equation $Z(t) = c_1 Z_1(t) + c_2 Z_2(t) + \ldots + c_s Z_s(t)$. From (12) it follows immediately that

$$\lim_{t \to \infty} Z(t) = \frac{1}{\mu} \sum_{k=1}^{s} c_k h_k = \frac{1}{\mu} \int_0^\infty z(t) \mathrm{d}t. \tag{13}$$

If we consider special intervals of form $[a_k, b_k) = [(k-1)h, kh)$, $k = 1, 2, \ldots$ and if

$$\sum_{k=1}^{\infty} |c_k| h = \int_0^\infty |z(t)| \, \mathrm{d}t < \infty, \tag{14}$$

then we conclude from

$$Z_k(t) = U(t - (k-1)h) - U(t - kh) \leqslant 1 + U(h)$$

(by (2)) that the series $Z(t) = c_1 Z_1(t) + c_2 Z_2(t) + \ldots$ converges *uniformly* in t. It is therefore permitted to interchange limit and summation and one again obtains from (12)

$$\lim_{t \to \infty} Z(t) = \lim_{t \to \infty} \sum_{k=1}^{\infty} c_k Z_k(t) = \frac{1}{\mu} \sum_{k=1}^{\infty} c_k h = \frac{1}{\mu} \int_0^\infty z(t) \mathrm{d}t. \tag{15}$$

This result can be generalized to a certain class of functions which can be approximated by step functions of the above kind. First let $z(t)$, $t > 0$, be an arbitrary bounded function and let, for $k = 1, 2, \ldots$,

$$\underline{m}_k = \inf_{(k-1)h \leqslant x < kh} z(x), \quad \bar{m}_k = \sup_{(k-1)h \leqslant x < kh} z(x). \tag{16}$$

The reader is reminded that $\underline{m}_k$ and $\bar{m}_k$ depend on h.

$$\underline{\sigma} = h \sum_{k=1}^{\infty} \underline{m}_k, \quad \bar{\sigma} = h \sum_{k=1}^{\infty} \bar{m}_k \tag{17}$$

are called *lower* and *upper sums*. Such sums are considered when defining Riemann integrals, but only for *finite* integration intervals, so that only finite sums, but no series, appear in (17). If $\underline{\sigma}$ and $\bar{\sigma}$ converge for $h \to 0$ to the same limit σ, we call this limit the Riemann integral over the underlying finite interval. If the series $\underline{\sigma}$ and $\bar{\sigma}$ (17) converge for all $h > 0$ *absolutely* and if we have in addition

$$-\infty < \lim_{h \to 0} \underline{\sigma} = \lim_{h \to 0} \bar{\sigma} < \infty, \tag{18}$$

we call $z(t)$ *directly Riemann integrable* and, analogous to the usual Riemann integral, the common limit (18) is written as

$$\lim_{h \to 0} \underline{\sigma} = \lim_{h \to 0} \bar{\sigma} = \int_0^\infty z(t)\mathrm{d}t. \tag{19}$$

In the usual theory of the Riemann integral the *improper* Riemann integral is *not* defined by (19) but as the limit of proper Riemann integrals by

$$\int_0^\infty z(t)\mathrm{d}t = \lim_{a \to \infty} \int_0^a z(t)\mathrm{d}t \tag{20}$$

(if this limit exists).

Since in (19) this detour via the finite integration intervals is avoided one speaks of direct integration there. *The two definitions (19) and (20) are not equivalent.* The reason is that there are functions $z(t)$ for which the infinite integral (20) exists and is finite without (18) being satisfied, so that they are not directly Riemann integrable. Nevertheless, we have the following lemma:

Lemma

If $z(t)$ is *directly Riemann integrable* and if the improper integral (20) exists, the two integrals (19) and (20) are *identical*.

Proof. For a fixed $h > 0$, $\underline{\sigma}$ and $\bar{\sigma}$ in (17) are improper integrals in the sense of (20) for the corresponding step functions with the values $\underline{m}_k$ or $\bar{m}_k$ in the intervals $[(k-1)h, kh)$. For every fixed h we have

$$\underline{\sigma} \leqslant \int_0^\infty z(t)\mathrm{d}t \leqslant \bar{\sigma},$$

where the integral here is to be understood in the sense of (20). The assertion follows from this and from (19) with $h \to 0$. ∎

Equation (15) can now be generalized to the class of the directly Riemann integrable functions.

Theorem 3

If $z(t)$ is bounded and *directly Riemann integrable*, the solution $Z(t)$ (2.1.10) of the renewal equation (2.1.8) for a non-arithmetic renewal distribution F satisfies

$$\lim_{t \to \infty} Z(t) = \frac{1}{\mu} \lim_{h \to 0} \bar{\sigma} = \frac{1}{\mu} \lim_{h \to 0} \underline{\sigma} = \frac{1}{\mu} \int_0^\infty z(t)\mathrm{d}t. \tag{21}$$

Proof. We define $z_k(t) = 1$ for $(k-1)h \leqslant t < kh$, $z_k(t) = 0$ otherwise, and introduce the functions

$$\underline{z}(t) = \sum_{k=1}^{\infty} \underline{m}_k z_k(t), \quad \bar{z}(t) = \sum_{k=1}^{\infty} \bar{m}_k z_k(t).$$

Furthermore, let $Z_k(t)$ be the solution (2.1.10) of the renewal equation (2.1.8) for the function $z_k(t)$. The solutions (2.1.10) of (2.1.8) for the functions $\underline{z}(t)$ and $\bar{z}(t)$ are

$$\underline{Z}(t) = \sum_{k=1}^{\infty} \underline{m}_k Z_k(t), \quad \bar{Z}(t) = \sum_{k=1}^{\infty} \bar{m}_k Z_k(t).$$

It follows from (15) that

$$\lim_{t \to \infty} \underline{Z}(t) = \frac{1}{\mu} \underline{\sigma}, \quad \lim_{t \to \infty} \bar{Z}(t) = \frac{1}{\mu} \bar{\sigma},$$

since $\underline{\sigma}$ and $\bar{\sigma}$, according to our assumptions, converge absolutely for every $h > 0$ so that condition (14) is satisfied. From (2.1.10) it follows that $\underline{Z}(t) \leqslant Z(t) \leqslant \bar{Z}(t)$ for every $t > 0$. Thus $\underline{\sigma}/\mu \leqslant \lim \inf_{t \to \infty} Z(t) \leqslant \lim \sup_{t \to \infty} Z(t) \leqslant \bar{\sigma}/\mu$. The statement follows from this by letting $h \to 0$. ∎

When applying this fundamental theorem of renewal theory $z(t)$ is often a monotonic function. The following theorem specializes Theorem 3 for this case.

Theorem 4

If $z(t)$ is *monotonic*, bounded and if

$$\int_0^{\infty} |z(t)| dt < \infty,$$

then the solution $Z(t)$ (2.1.10) of the renewal equation for a *non-arithmetic* renewal distribution F satisfies

$$\lim_{t \to \infty} Z(t) = \frac{1}{\mu} \int_0^{\infty} z(t) dt. \tag{22}$$

Proof. From the assumption it follows that either $z(t) \geqslant 0$ and monotonically non-increasing, or $z(t) \leqslant 0$ and monotonically non-decreasing, and that $z(t) \to 0$ as $t \to \infty$. It is sufficient just to look at the first case $(z(t) \geqslant 0)$ since the other one follows in a similar way. For $h > 0$ $\bar{m}_k = z((k-1)h) \geqslant 0$ and $\underline{m}_k = z(kh) \geqslant 0$. Hence

$$\underline{\sigma} = h \sum_{k=1}^{\infty} z(kh) \leqslant \int_0^{\infty} z(t) dt < \infty$$

and $\underline{\sigma}$ converges absolutely. Since $\bar{\sigma} = \underline{\sigma} + hz(0)$ we have $\bar{\sigma} - \underline{\sigma} = hz(0) \to 0$ as $h \to 0$. Furthermore we have for every $h > 0$

$$\underline{g} \leqslant \int_0^\infty z(t)\,\mathrm{d}t \leqslant \bar{o}.$$

hence

$$\lim_{h\to 0} \underline{g} = \lim_{h\to 0} \bar{o} = \int_0^\infty z(t)\,\mathrm{d}t < \infty.$$

$z(t)$ is thus directly Riemann integrable and the statement follows from Theorem 3. ∎

Examples

(b) *Residual life, age, inspected life*. In the renewal equation of Example (2.1.b) for the distribution $R_z(t)$ of the residual life the function $z(t) = 1 - F(t + z)$ is *monotonically decreasing* in t. If the expected value μ of the renewal distribution F is finite, it follows that (see (1.3.38))

$$\int_0^\infty (1 - F(t + z))\,\mathrm{d}t = \int_z^\infty (1 - F(t))\,\mathrm{d}t \leqslant \mu < \infty.$$

Thus the conditions of Theorem 4 are satisfied and

$$\lim_{t\to\infty} R_z(t) = R_z(\infty) = \frac{1}{\mu}\int_z^\infty (1 - F(t))\,\mathrm{d}t, \quad z > 0.$$

This is the limit distribution for the residual life. This limit distribution obviously has the density function $(1 - F(x))/\mu$, $x > 0$: compare this result with the discrete case of Example (2.2.c). The same limit distribution is obtained for the age distribution $A_z(t)$. Theorem 4 can equally be applied to the distribution $L_z(t)$ of the inspected life: we can show by partial integration (see (1.3.36)) that

$$\lim_{t\to\infty} L_z(t) = L_z(\infty) = \frac{1}{\mu}\int_0^\infty (1 - F(\max(z, t)))\,\mathrm{d}t$$

$$= \frac{1}{\mu}\left\{ \int_0^z (1 - F(z))\,\mathrm{d}t + \int_z^\infty (1 - F(t))\,\mathrm{d}t \right\}$$

$$= \frac{1}{\mu}\left\{ (1 - F(z))z + \mu - \int_0^z (1 - F(t))\,\mathrm{d}t \right\} = \frac{1}{\mu}\int_z^\infty t\,\mathrm{d}F(t).$$

The expected value of the limit distribution for the inspected life is

$$\eta = \frac{1}{\mu}\int_0^\infty t^2\,\mathrm{d}F(t) \geqslant \mu$$

as we can see from the inequality $E[X^2] \geqslant (E[X])^2$ (see (1.3.35)). This confirms the remark made in Example (2.1.b) that with an arbitrary inspection time t one tends to encounter longer rather than shorter periods of life.

For the *Poisson process* (see Example (2.1.*a*)) we obtain $R_z(\infty) = A_z(\infty) = e^{-\lambda z}$, i.e. the limit distributions of the residual life and of the age are equal to the renewal distribution and $L_z(\infty) = (1 + \lambda z)e^{-\lambda z}$, and it follows in this particular case that $\eta = 2\mu$, i.e. that the expected value of the inspected life is twice as big as the expected length of the renewal intervals.   □

(*c*) *Age-renewal policies.* For the renewal equation of the renewal cost in Example (2.1.*c*) Theorem 3 cannot be applied directly if there is no discount, i.e. $\rho = 0$, since the function $z(t)$ is not integrable. It is also obvious that $c(t)$ cannot tend to a finite limit as $t \to \infty$. This is confirmed by the following considerations.

Let $\bar{U}(t)$ be the renewal function for the renewal distribution $\bar{F}(x)$ (Example (2.1.*c*)) and $d(t) = z(T)(\bar{U}(t) + 1) - c(t)$. If we compare the renewal equations for $\bar{U}(t)$ (2.1.9) and for $c(t)$ (Example (2.1.*c*)), we realize that $d(t)$ satisfies the renewal equation

$$d(t) = (z(T) - z(t)) + \int_0^t d(t-x)\,\mathrm{d}\bar{F}(x).$$

Here the function $z(T) - z(t)$ decreases monotonically.

$$z(T) - z(t) = \begin{cases} c_1(F(T-0) - F(t)) + c_2(1 - F(T-0)), & t < T, \\ 0, & t \geqslant T. \end{cases}$$

Furthermore, for $T < \infty$,

$$\int_0^\infty (z(T) - z(t))\,\mathrm{d}t = T(c_1 F(t-0) + c_2(1 - F(T-0))) - \int_0^T c_1 F(t)\,\mathrm{d}t < \infty.$$

For $T = \infty$ this integral becomes equal to $c_1\mu$ (see the formula for the expected value (1.3.38)), where μ is the expected value of $F(x)$. Theorem 4 can be applied to $d(t)$ and yields for $T < \infty$

$$\lim_{t \to \infty} d(t) = \frac{1}{\bar{\mu}}\left\{ T(c_1 F(T-0) + c_2(1 - F(t-0))) - \int_0^T c_1 F(t)\,\mathrm{d}t \right\},$$

where $\bar{\mu}$ is the expected value of the distribution $\bar{F}(x)$. From this and from the elementary renewal theorem (the corollary to Theorem 2), it follows that

$$\lim_{t \to \infty} \frac{c(t)}{t} = \lim_{t \to \infty} \frac{z(T)(\bar{U}(t) + 1)}{t} = \frac{z(T)}{\bar{\mu}} = \frac{c_1 F(T-0) + c_2(1 - F(T-0))}{\int_0^T x\,\mathrm{d}\bar{F}(x)}.$$

This expected average cost per unit time was already determined in Example (*a*) in a different way. It follows similarly for $T = \infty$ that this average cost is equal to c_1/μ. In the case of cost consideration without discount we can again determine the optimal renewal age for any given distribution $F(x)$ which will minimize the above average cost.

If we set $c_1 = 0$ and $c_2 = 1$, then $c(t)$ represents the expected value of the number of planned age renewals up to t. By the above result we can expect on average $(1 - F(t - 0))/\bar{\mu}$ age renewals per unit time. Similarly we find with $c_1 = 1$ and $c_2 = 0$ that on average $F(T - 0)/\bar{\mu}$ failures (unplanned renewals) per unit time are to be expected. □

The result for the average cost in the above Example (c) represents only a special case of a more general result. Suppose $\{(Y_i, X_i), i = 1, 2, \ldots\}$ is a sequence of mutually *independent* and *equally distributed* pairs of random variables. Note that Y_i and X_i do *not* have to be mutually independent for a fixed i; X_i is to take only *positive* values and is to have a distribution function $F(x)$ with $F(0) = 0$; Y_i can be arbitrarily distributed. The sequence of the X_i defines a *renewal process* and we let $N(t), t \geq 0$, be the associated counting process (see (2.1.3)). The stochastic process $\{Z(t), t \geq 0\}$ defined by

$$Z(t) = \sum_{i=1}^{N(t)+1} Y_i \tag{23}$$

is called a *cumulative process*. If we consider Y_i as a pay-off at the beginning of the ith renewal interval, then $Z(t)$ is the sum of the pay-offs up to time t. We have the following result.

Theorem 5

If the Y_i have a *finite* expected value $E[Y_i] = E[Y_1]$ and if the renewal process of the X_i is *recurrent* we have

$$\lim_{t \to \infty} \frac{E[Z(t)]}{t} = \frac{E[Y_1]}{E[X_1]}. \tag{24}$$

Proof. Let $A(t) = E[Z(t)]$; i.e. $A(t)$ is the expected value of the pay-offs up to time t. At $t = 0$ we already have the expected pay-off $E[Y_1]$. If the next renewal takes place at $x < t$, we can expect in addition the pay-off $A(t - x)$ up to t. Hence

$$A(t) = E[Y_1] + \int_0^t A(t - x)\,dF(x). \tag{25}$$

If $U(t)$ is the renewal function for the renewal process defined by the X_i it follows from Theorem 2.1.2 that $A(t) = E[Y_1](1 + U(t))$ and the elementary renewal theorem (11) implies our assertion. ∎

Note especially that this theorem holds independently of whether the renewal distribution F of the X_i is arithmetic or not. In the *arithmetic* case it is usually more natural to consider the cumulative processes only for integral times $t = n = 0, 1, 2, \ldots$. Between these times nothing can change anyway. In this case we have

$$\lim_{n \to \infty} \frac{E[Z(n)]}{n} = \frac{E[Y_1]}{E[X_1]}. \tag{26}$$

In the above Example (c) the Y_i are the costs during the ith renewal period and $E[Y_i] = c_1 F(T-0) + c_2(1 - F(T-0))$. In this example X_i has the distribution function $\bar{F}(x)$. The result for $\lim c(t)/t$ matches (24) precisely.

Notes to Chapter 2

Renewal theory, as it is developed here, goes back to *W. Feller* (see [1.3]). We refer the reader to Vol. 2 of that book for the proof of the two lemmas in the demonstration of Theorem 2.3.2. [8] is an older account of renewal theory. [6] is a more recent discussion of the mathematical foundation of renewal processes and generalizations of them. There are a number of monographs: [1], [2], [4], [5], [7], [9], [10] showing the application of renewal theory to replacement and maintenance problems. [3] contains a systematic study of various preventive maintenance policies of the kind introduced in Example (2.1.c).

Renewal theory has its most important applications in the theory of Markov chains, Markov renewal processes, queueing systems, simulation and Monte-Carlo methods. The reader is referred to the chapters which follow.

3
Markov chains

3.1. Markov chains with discrete time parameter

A sequence $\{X_n, n = 0, 1, 2, \ldots\}$ of random variables is called a *stochastic process with discrete time parameter*, or simply *with discrete parameter* (see Section 1.2). We assume that all variables of the sequence have the same value space I. In the following I is assumed to be *countable* so that the variables X_n are *discrete* (see Section 1.2). This includes, of course, the possibility of finite sets I. The set I is called the *state space* and every element of I is called a *state* of the stochastic process. If the random variable X_n takes the value $i \in I$ we say that the process is in state i at time n.

A stochastic process with discrete parameter and a countable state space is called a *Markov chain with discrete parameter* if the process has the following property: for all n-tuples $0 \leqslant t_1 < \ldots < t_n$ of values of the time parameter and $i_1, \ldots, i_n$ of states, $n \geqslant 2$, the conditional probabilities of $X_{t_n} = i_n$, given that $X_{t_1} = i_1, \ldots, X_{t_{n-1}} = i_{n-1}$, is described by

$$P[X_{t_n} = i_n | X_{t_1} = i_1, \ldots, X_{t_{n-1}} = i_{n-1}] = P[X_{t_n} = i_n | X_{t_{n-1}} = i_{n-1}]. \qquad (1)$$

This is the so-called *Markov property*. It implies in particular that

$$P[X_n = i_n | X_0 = i_0, \ldots, X_{n-1} = i_{n-1}] = P[X_n = i_n | X_{n-1} = i_{n-1}],$$

and thus Markov chains have the property that at every step the probability of the new state *depends only on the last state but not on the previous states*. It is this property which makes the Markov chains into such useful models for real processes. More generally we can say that for Markov chains the *future* development is determined by the *present* state alone and does not depend on *the past*. This statement is made precise in the following assertion:

Theorem 1
If $0 \leqslant t_1 < \ldots < t_n < \ldots < t_{n+m}$ are values of the time parameter and $i_1, \ldots, i_n, \ldots, i_{n+m}$ are the associated states, where $n \geqslant 2, m \geqslant 0$, we have

$$P[X_{t_\nu} = i_\nu; n \leqslant \nu \leqslant n + m | X_{t_\nu} = i_\nu; 1 \leqslant \nu \leqslant n - 1]$$
$$= P[X_{t_\nu} = i_\nu; n \leqslant \nu \leqslant n + m | X_{t_{n-1}} = i_{n-1}]. \qquad (2)$$

The probabilities of future states depend only on the *last* of the observed states and do not depend on the previously observed states.

Proof. We prove this by induction on m. For $m = 0$ (2) is the Markov property (1). Suppose the assertion holds for m. Then it follows by using (in this order) (1.1.19), the induction hypothesis (2) for m, (1), once again (1) and (1.1.19):

$$P[X_{t_v} = i_v; n \leqslant v \leqslant n + m + 1 | X_{t_v} = i_v; 1 \leqslant v \leqslant n - 1]$$
$$= P[X_{t_v} = i_v; n \leqslant v \leqslant n + m | X_{t_v} = i_v; 1 \leqslant v \leqslant n - 1]$$
$$P[X_{t_{n+m+1}} = i_{n+m+1} | X_{t_v} = i_v; 1 \leqslant v \leqslant n + m]$$
$$= P[X_{t_v} = i_v; n \leqslant v \leqslant n + m | X_{t_{n-1}} = i_{n-1}]$$
$$P[X_{t_{n+m+1}} = i_{n+m+1} | X_{t_{n+m}} = i_{n+m}$$
$$= P[X_{t_v} = i_v; n \leqslant v \leqslant n + m | X_{t_{n-1}} = i_{n-1}]$$
$$P[X_{t_{n+m+1}} = i_{n+m+1} | X_{t_v} = i_v; n - 1 \leqslant v \leqslant n + m]$$
$$= P[X_{t_v} = i_v; n \leqslant v \leqslant n + m + 1 | X_{t_{n-1}} = i_{n-1}].$$

This result therefore also holds for $m + 1$. ∎

The probabilistic behaviour of a stochastic process with discrete parameter and countable state space is normally described and determined by the *finite-dimensional distributions* $P[X_{t_0} = i_0, \ldots, X_{t_n} = i_n], 0 \leqslant t_0 < \ldots < t_n, n \geqslant 0$ (see Section 1.2). For these we have in general (see Example (1.2.h))

$$P[X_{t_0} = i_0, \ldots, X_{t_n} = i_n]$$
$$= P[X_{t_0} = i_0] \prod_{v=1}^{n} P[X_{t_v} = i_v | X_{t_\lambda} = i_\lambda; 0 \leqslant \lambda \leqslant v - 1].$$

For Markov chains it now follows from the Markov property (1) that

$$P[X_{t_0} = i_0, \ldots, X_{t_n} = i_n] = P[X_{t_0} = i_0] \prod_{v=1}^{n} P[X_{t_v} = i_v | X_{t_{v-1}} = i_{v-1}]. \quad (3)$$

In particular we have

$$P[X_0 = i_0, X_1 = i_1, \ldots, X_n = i_n] = P[X_0 = i_0] \prod_{t=1}^{n} P[X_t = i_t | X_{t-1} = i_{t-1}]. \quad (4)$$

$p_i = P[X_0 = i]$ is the *initial distribution* of the Markov chain. The $P[X_t = j | X_{t-1} = i]$ are called *transition probabilities* of the Markov chain. They depend on the states i, j and, in general, also on t, and they express the probability for the event that the process at time t is in the state j if at time $t - 1$ it is in the state i.

If the transition probabilities do not depend on t we can write $P[X_t = j \mid X_{t-1} = i] = p_{ij}$. In this case we speak of a Markov chain with *stationary transition probabilities* or of a *homogeneous* Markov chain. In most applications we deal with homogeneous chains; see the examples below. If the transition probabilities depend on t, we speak of an *inhomogeneous Markov chain.*

For a *homogeneous* Markov chain (4) takes the form

$$P[X_0 = i_0, X_1 = i_1, \ldots, X_n = i_n] = p_{i_0} \prod_{t=1}^{n} p_{i_{t-1} i_t}. \tag{5}$$

Hence, for the computation of the distributions $P[X_0 = i_0, \ldots, X_n = i_n]$ it is sufficient to know the initial distribution p_i and the transition probabilities p_{ij}. Below we show that these data also allow the computation of the more general distributions in (3).

This means that it is possible to determine a homogeneous Markov chain completely by stating the initial distribution p_i and the transition probabilities p_{ij}. This is, in fact, how Markov chains are introduced as models in operations research as well as in other applications. The p_i and p_{ij} form the *data* for the model. For the initial distribution we must have

$$p_i \geqslant 0 \text{ for all } i \in I; \quad \sum_{i \in I} p_i = 1,$$

the usual condition for a discrete distribution. Similarly the p_{ij} form a discrete distribution for every fixed $i \in I$, so that

$$p_{ij} \geqslant 0 \text{ for all } i, j \in I, \quad \sum_{j \in I} p_{ij} = 1 \text{ for all } i \in I. \tag{6}$$

The transition probabilities p_{ij} can be combined in the matrix $P = (p_{ij})$, the *transition matrix.* A matrix P whose entries satisfy the conditions (6) is called *stochastic.* In introducing a Markov chain it is often sufficient to specify the transition matrix P and to leave the initial distribution open. Although the Markov chain is not completely determined by this we shall conform to the general convention and speak of a Markov chain with transition matrix P.

Examples

(a) *Markov chains with two states.* The simplest non-trivial case of a Markov chain is the one with two states, $I = \{1, 2\}$. The transition matrix is of the form

$$P = \begin{bmatrix} a & 1-a \\ 1-b & b \end{bmatrix}. \qquad \square$$

(b) *Discrete renewal process.* In connection with the discrete renewal processes presented in Chapter 2 various Markov chains can be introduced. The most interesting one is connected with the *age* of the renewal process at time n (see Example (2.2.c)). The state of the system at time n is equal to the age of the renewal process at time n, where, as in Example (2.2.c), the age at time n is always determined *before* a possible renewal. Hence $I = \{1, 2, \ldots\}$. From one time interval to the next, transitions are only possible from i to $i + 1$ and to 1 so that $p_{ij} = 0$ for $j \neq i + 1$ and $j \neq 1$. p_{ii+1} is the conditional probability of reaching age $i + 1$ given that age i has been reached. Note that with the above convention of observation *before* a possible renewal, reaching age $i + 1$ means that no renewal has taken place during the first i time intervals after the last renewal. Hence, by the definition of conditional probabilities (1.1.10)

$$p_{ii+1} = \left(1 - \sum_{k=1}^{i} f_k\right)\Big/\left(1 - \sum_{k=1}^{i-1} f_k\right), \quad i \geq 1$$

and

$$p_{i1} = f_i\Big/\left(1 - \sum_{k=1}^{i-1} f_k\right), \quad i \geq 1,$$

where $\{f_k, k = 1, 2, \ldots\}$, represents the renewal distribution as in Example (2.2.c).

A second Markov chain can be introduced in connection with the *residual life* (see Example (2.2.c)). The state of the system at time n is now the residual life at time n, which, in contrast to age, is always observed *after* a possible renewal. Therefore $I = \{1, 2, \ldots\}$. Transitions from $i = 1$ to $j = 1, 2, \ldots$ are possible with transition probabilities $p_{1j} = f_j, j = 1, 2, \ldots$. From $i \neq 1$ transitions are only possible to $i - 1$ so that $p_{ii-1} = 1$ and $p_{ij} = 0$ for $j \neq i - 1, i > 1$. $\qquad\square$

(c) *Renewal policy.* In Example (2.1.c) we introduced the notion of age-renewal policy, although for non-discrete renewal processes. This notion can easily be applied to discrete processes. Let the renewal age be denoted by T. Let the policy, however, be modified in such a way that when a renewal takes place, not a new plant or machine is acquired, but one of age $k, 0 < k < T$ (a special offer). Let the state of the system at time n be, as in Example (b), the age of the system *before* a possible renewal at that moment. At the beginning the system may be of any age so that $I = \{1, 2, \ldots\}$. The appropriate modification of Example (b) results in the following transition probabilities:

$$p_{ii+1} = \left(1 - \sum_{k=1}^{i} f_k\right)\Big/\left(1 - \sum_{k=1}^{i-1} f_k\right), \quad i \neq T,$$

$$p_{ik+1} = f_i\Big/\left(1 - \sum_{k=1}^{i-1} f_k\right), \quad i \neq T,$$

$$p_{Tk+1} = 1.$$

All other transition probabilities are zero. $\qquad\square$

(*d*) *Random walks.* Numbering states by $1, 2, \ldots$ is not in every case the most suitable and natural notation. If, for example, we look at a particle which can jump on the real line in one step from any integer i to any adjacent integer $i - 1$ or $i + 1$, we naturally choose $I = \{\ldots, -1, 0, +1, +2, \ldots\}$. Let the transition probabilities be $p_{ii-1} = p$, $p_{ii+1} = q = 1 - p$ and $p_{ij} = 0$, if $j \neq i - 1$ and $j \neq i + 1$. The Markov chain thus defined is called an *unbounded random walk.* If we modify an unbounded random walk in such a way that $p_{00} = 1$, $p_{0j} = 0$ for $j \neq 0$ we obtain a *random walk with a one-sided absorbing boundary.* For reasons of symmetry we can limit ourselves in this case to the consideration of the state space $I = \{0, 1, 2, \ldots\}$. If we further put, for $k > 0$, $p_{kk} = 1$, $p_{kj} = 0$ for $j \neq k$ and if we limit our consideration to the finite state space $I = \{0, 1, 2, \ldots, k\}$, we call this a *random walk with a two-sided absorbing boundary.* By varying the behaviour at the boundary we can introduce many more variants of these random walks. □

(*e*) *Branching processes.* Let us look at a family of individuals or elements of the same kind, each of which can have a certain number of offspring. Let the number of offspring for any one individual be a discrete random variable Z whose possible values are $0, 1, 2, \ldots$. We assume the random variables Z to be independent and identically distributed for all individuals. Let X_n be the number of individuals at time n, i.e. the size of the nth generation. We then have, if Z_k is the offspring of the kth individual

$$X_{n+1} = \sum_{k=1}^{X_n} Z_k.$$

$\{X_n, n = 0, 1, 2, \ldots\}$ forms a Markov chain whose transition probabilities are defined by

$$p_{ij} = P[X_{n+1} = j \mid X_n = i] = P\left[\sum_{k=1}^{i} Z_k = j\right].$$

A major problem in connection with such *branching processes* is the determination of the probability of *dying out*, i.e. the probability that for some n the population becomes $X_n = 0$. At this moment we can consider the process as discontinued since no further descendants can be produced and thus $X_{n+m} = 0$ for all $m \geqslant 0$. □

(*f*) *Stock-keeping.* Assume that the daily (weekly, monthly) demand for a commodity which is kept in stock is given by independent identically distributed random variables $Z_1, Z_2, \ldots$ with a known distribution. The Z_i can take the values $0, 1, 2, \ldots$. For the stocking-up policy of the warehouse let us assume the following rule: as soon as the stock is below a critical value s, it is immediately re-stocked to a maximum S. At the same time orders are dispatched which could not be dealt with from the existing stock. If X_n denotes the stock at the beginning of the nth day, then

$$X_{n+1} = \begin{cases} X_n - Z_n & \text{as long as } X_n - Z_n > s, \\ S & \text{as long as } X_n - Z_n \leqslant s. \end{cases}$$

$\{X_n, n = 0, 1, 2, \ldots\}$ forms a Markov chain with the following transition probabilities for $i = s + 1, \ldots, S$:

$$p_{ij} = P[X_{n+1} = j | X_n = i] = P[Z_n = i - j], \quad s < j \leqslant i < S \text{ or } s < j < i = S,$$

$$p_{iS} = P[Z_n \geqslant i - s], \quad s < i < S,$$

$$p_{SS} = P[Z_n = 0] + P[Z_n \geqslant S - s].$$

All other transition probabilities are equal to zero.   □

The probabilities $P[X_{t+n} = j | X_t = i]$ are called *n-step transition probabilities*. For the *n*-step transition probabilities there exists a fundamental formula which is proved in the following theorem.

Theorem 2
For all $t = 0, 1, 2, \ldots$ and $n, m = 1, 2, \ldots$

$$P[X_{t+n+m} = j | X_t = i] = \sum_{k \in I} P[X_{t+n+m} = j | X_{t+n} = k] P[X_{t+n} = k | X_t = i].$$
(7)

For *homogeneous* Markov chains the *n*-step transition probabilities $P[X_{t+n} = j | X_t = i]$ do not depend on t and we can therefore write $P[X_{t+n} = j | X_t = i] = p_{ij}^{(n)}$. In this case (7) reads

$$p_{ij}^{(n+m)} = \sum_{k \in I} p_{ik}^{(n)} p_{kj}^{(m)}.$$
(8)

Equations (7) and (8) are the so-called *Chapman–Kolmogorov equations* for inhomogeneous and homogeneous Markov chains, respectively.

Proof. All summations in the following are taken over I. By using the Markov property (1) as well as (1.1.15) and (1.1.19) it follows that

$$P[X_{t+n+m} = j | X_t = i]$$

$$= \sum_k P[X_{t+n+m} = j, X_{t+n} = k | X_t = i]$$

$$= \sum_k P[X_{t+n+m} = j | X_{t+n} = k, X_t = i] P[X_{t+n} = k | X_t = i]$$

$$= \sum_k P[X_{t+n+m} = j | X_{t+n} = k] P[X_{t+n} = k | X_t = i].$$
(9)

This formula has the following interpretation: if the chain changes from i

at time t to j at time $t + n + m$, then at time $t + n$ it must be in a state $k = 1, 2, \ldots$. The probability that the chain goes from i via k to j is equal to $P[X_{t+n+m} = j \mid X_{t+n} = k] P[X_{t+n} = k \mid X_t = i]$, and we only have to sum up all the possibilities $k = 1, 2, \ldots$ in order to obtain $P[X_{t+n+m} = j \mid X_t = i]$.

If we put $m = 1$ in (9) we see by induction on n that, for *homogeneous* chains, $P[X_{t+n} = j \mid X_t = i] = p_{ij}^{(n)}$ does not depend on t. ■

For the rest of this section as well as for the following we concentrate mostly on the consideration of *homogeneous* Markov chains. In this case it follows from (8) that

$$p_{ij}^{(n)} = \sum_k p_{ik}^{(n-1)} p_{kj} \tag{10}$$

represents a recursive formula for the computation of the $p_{ij}^{(n)}$ from p_{ij}. We introduce the notion of the *absolute distribution* $p_i^{(n)} = P[X_n = i]$, where $p_i^{(0)} = p_i$ is the initial distribution. By (3) and (1.1.9)

$$p_i^{(n)} = \sum_k P[X_n = i; X_0 = k] = \sum_k P[X_0 = k] P[X_n = i \mid X_0 = k]$$

$$= \sum_k p_k p_{ki}^{(n)}. \tag{11}$$

Similarly we obtain

$$p_i^{(n)} = \sum_k p_k^{(n-1)} p_{ki}. \tag{12}$$

Formula (3) can now be written in the form

$$P[X_{t_0} = i_0, \ldots, X_{t_n} = i_n] = p_{i_0}^{(t_0)} \prod_{\nu=1}^{n} p_{i_{\nu-1} i_\nu}^{(t_\nu - t_{\nu-1})}. \tag{13}$$

Since the n-step transition probabilities for homogeneous chains are, according to (10), determined by the p_{ij}, we can see that the finite-dimensional distributions (3) can also be determined by p_i and p_{ij}.

Many of these relations can be expressed concisely in matrix notation. Let $P^{(n)}$ be the matrix of the n-step transition probabilities $p_{ij}^{(n)}$, $P^{(1)} = P$ the transition matrix, $p^{(n)}$ the row vector of the probabilities of the absolute distributions $p_i^{(n)}$. Then relation (10) in matrix notation reads as $P^{(n)} = P^{(n-1)} P$, which implies that $P^{(2)} = P^{(1)} P = PP = P^2$ and hence $P^{(3)} = P^{(2)} P = P^2 P = P^3$. By induction on n we finally obtain $P^{(n)} = P^n$. The Chapman–Kolmogorov equations (8) appear in this notation as $P^{(n+m)} = P^{(n)} P^{(m)}$ and express, as we can see now, nothing more than $P^{n+m} = P^n P^m = P^m P^n$. For (11) and (12) we can now write $p^{(n)} = p^{(0)} P^n$ and $p^{(n)} = p^{(n-1)} P$, respectively.

Example

(g) *Computation of n-step transition probabilities for chains with finite state space.* Formula (10) permits the computation of n-step transition probabilities. With the example of a chain with two states we shall show how n-step transition probabilities of chains with *finite* state spaces can also be computed with the help of generating functions. The transition matrix is given in Example (*a*). Let $P_{ij}(z)$ be the generating functions of the sequences $\{p_{ij}^{(n)}, n = 1, 2, \ldots\}$

$$P_{ij}(z) = \sum_{n=1}^{\infty} p_{ij}^{(n)} z^{n-1},$$

see (2.2.1). If we multiply both sides by $z p_{ki}$ and sum over i, we obtain

$$z \sum_i p_{ki} P_{ij}(z) = P_{kj}(z) - p_{kj}.$$

If we fix j we obtain a system of linear equations for $P_{1j}(z), P_{2j}(z), \ldots$, which we can solve appropriately. If P is the transition matrix and $P(z)$ the matrix of the $P_{ij}(z)$, the equation system is $zPP(z) = P(z) - P$ and the solution is $P(z) = (I - zP)^{-1}P$. The elements of $(I - zP)^{-1}$ are rational functions in z whose denominators are equal to the determinant of $(I - zP)$. For

$$P = \begin{bmatrix} \frac{1}{2} & \frac{1}{2} \\ \frac{1}{4} & \frac{3}{4} \end{bmatrix},$$

for example, the determinant of $I - zP$ is equal to $(1 - \frac{1}{2}z)(1 - \frac{3}{4}z) - \frac{1}{8}z^2$. The quadratic function has zeros at $z = 1$ and $z = 4$ and can therefore also be written in the form $(1 - z)(1 - \frac{1}{4}z)$. We thus have

$$(I - zP)^{-1} = \frac{1}{(1-z)(1-\frac{1}{4}z)} \begin{bmatrix} 1 - \frac{3}{4}z & \frac{1}{2}z \\ \frac{1}{4}z & 1 - \frac{1}{2}z \end{bmatrix}.$$

If we apply the well-known decomposition of rational functions into partial fractions to the elements of this matrix, we find that

$$\frac{1 - \frac{3}{4}z}{(1-z)(1-\frac{1}{4}z)} = \frac{\frac{1}{3}}{1-z} + \frac{\frac{2}{3}}{1-\frac{1}{4}z},$$

$$\frac{\frac{1}{2}z}{(1-z)(1-\frac{1}{4}z)} = \frac{\frac{2}{3}}{1-z} - \frac{\frac{2}{3}}{1-\frac{1}{4}z},$$

$$\frac{\frac{1}{4}z}{(1-z)(1-\frac{1}{4}z)} = \frac{\frac{1}{3}}{1-z} - \frac{\frac{1}{3}}{1-\frac{1}{4}z},$$

$$\frac{1 - \frac{1}{2}z}{(1-z)(1-\frac{1}{4}z)} = \frac{\frac{2}{3}}{1-z} + \frac{\frac{1}{3}}{1-\frac{1}{4}z},$$

and therefore

$$(I - zP)^{-1} = \frac{1}{1-z} \begin{bmatrix} \frac{1}{3} & \frac{2}{3} \\ \frac{1}{3} & \frac{2}{3} \end{bmatrix} + \frac{1}{1 - \frac{1}{4}z} \begin{bmatrix} \frac{2}{3} & -\frac{2}{3} \\ -\frac{1}{3} & \frac{1}{3} \end{bmatrix}.$$

If we multiply this expression by P and if we finally expand $1/(1-z)$ and $1/(1 - \frac{1}{4}z)$ as geometric series, we can see by comparison of coefficients with $P(z)$ that

$$P^{(n)} = P^n = \begin{bmatrix} \frac{1}{3} & \frac{2}{3} \\ \frac{1}{3} & \frac{2}{3} \end{bmatrix} + (\frac{1}{4})^{(n-1)} \begin{bmatrix} \frac{1}{6} & -\frac{1}{6} \\ -\frac{1}{12} & \frac{1}{12} \end{bmatrix}.$$

For $n \to \infty$, P^n tends to the first matrix of this expression. The question of convergence of P^n in the general case will be taken up in the next section. $\square$

The computation of n-step transition probabilities using (10) is much simplified by using a computer. The method introduced in this example has, however, the advantage that P^n is presented in a much simpler and more explicit form, and this may sometimes prove very useful.

3.2. Classification of states and asymptotic distributions

In this section we shall only consider homogeneous Markov chains. If the Markov chain is in state i, a state j at a later time can only be reached if there exists an integer $m > 0$ such that $p_{ij}^{(m)} > 0$. In this case we say that *state i leads to state j* and write $i \Rightarrow j$. If i leads to k and k to j, $i \Rightarrow k$, $k \Rightarrow j$, there exist two integers $m > 0$ and $n > 0$ such that $p_{ik}^{(m)} > 0$ and $p_{kj}^{(n)} > 0$. From the Chapman–Kolmogorov equations (3.1.8) it follows that $p_{ij}^{(m+n)} \geqslant p_{ik}^{(m)}p_{kj}^{(n)} > 0$. Therefore it follows from $i \Rightarrow k$, $k \Rightarrow j$ that $i \Rightarrow j$, i.e. the relation is *transitive*. If i leads to j and j to i, it means that the two states *communicate* and we write $i \Leftrightarrow j$. The relation $\Leftrightarrow$ is *transitive* and *symmetric* which means that $i \Leftrightarrow k$, $k \Leftrightarrow j$ imply $i \Leftrightarrow j$ and $i \Leftrightarrow j$ implies $j \Leftrightarrow i$. However, since a state i does not necessarily have to lead to itself, the relation is not necessarily *reflexive* for all states, i.e. $i \Leftrightarrow i$ does not necessarily hold.

If i does not lead to itself, i cannot communicate with any other state. For if there were a j such that $i \Leftrightarrow j$, then, for reasons of symmetry, it would follow that $j \Leftrightarrow i$ and from $i \Leftrightarrow j$ and $j \Leftrightarrow i$, together with transitivity, $i \Leftrightarrow i$ would follow. States which do not lead to themselves can only, so to speak, occur momentarily, at a certain moment, and never again afterwards.

The relation $\Leftrightarrow$ is reflexive, symmetric and transitive on the set of states which lead to themselves and, therefore leads to a *partition into classes* of the states which lead to themselves. All states which communicate among each other form a *class*. Each state which cannot lead to itself, is assumed to form a class of its own. With this convention the entire state space is decomposed into disjoint classes.

A set C of states is *closed* if no state in C leads to a state outside C. A class in the above sense may, but does not have to, be closed. However, no proper subset of a class can be closed. The closed classes form *minimal* closed sets, i.e. those which do not contain any smaller closed sets.

Theorem 1

A set of states C is closed if and only if the matrix of the transition probabilities (p_{ij}), $i \in C, j \in C$, is *stochastic*.

Proof. If C is closed we must have $p_{ij} = 0$ for all $i \in C$ and all $j \notin C$. For $i \in C$ we, therefore, have

$$\sum_j p_{ij} = \sum_{j \in C} p_{ij} = 1.$$

Conversely, if (p_{ij}), $i \in C, j \in C$, is stochastic, it follows from relation (3.1.10) that $p_{ij}^{(n)} = 0$ for all $n = 1, 2, \ldots$, if $i \in C$ and $j \notin C$. Hence $i \in C$ cannot lead to any state outside C, and C is closed. ∎

This theorem implies that every closed set by itself already forms a Markov chain. A Markov chain has in general several *closed* classes, and each of these classes determines by itself a Markov chain which cannot be reduced any further. By suitable re-arrangement of the rows and columns of the transition matrix P we can write it as

$$P = \begin{bmatrix} P_1 & 0 & \cdots & 0 & 0 \\ 0 & P_2 & \cdots & 0 & 0 \\ \cdot & & & \cdot & \cdot \\ \cdot & & & \cdot & \cdot \\ \cdot & & & \cdot & \cdot \\ 0 & 0 & \cdots & P_r & 0 \\ & & T & & \end{bmatrix},$$

where $P_1, \ldots, P_r$ are stochastic matrixes for r closed classes $C_1, \ldots, C_r$. The matrix P^n has the same structure with $P_1^n, \ldots, P_r^n$ instead of $P_1, \ldots, P_r$ and a matrix $T^{(n)}$ instead of T.

The state space I is always a closed set. If the state space I is furthermore a closed *class*, the Markov chain is called *irreducible*. In an irreducible chain every state communicates with every other state and the transition matrix contains no stochastic submatrix.

If a state $i \in I$, which leads to itself, is a closed class in itself, i.e. if it does not lead to a state $j \neq i$, then we must have $p_{ii} = 1$ (see also Theorem 1). Conversely, if for a state i, $p_{ii} = 1$, then it forms by itself a closed class. Such a state is called *absorbing*.

Examples

(a) *Discrete renewal process.* In the Markov chain, Example (3.1.*b*), which is connected with age, as well as in the one connected with the remaining life of a renewal process, all states communicate with each other. If, however, there exists a *maximal* renewal interval m, such that $f_i = 0$ for $i > m$, this is no longer true for the state space $I = \{1, 2, \ldots \}$. It is in this case natural to restrict the state space I to the set $\{1, 2, \ldots, m\}$. The chains are then always *irreducible*. In contrast, for a similar chain, of the renewal policy in Example (3.1.*c*), the states $i \leqslant k$ and $i > T$ do not lead to themselves while the states $k < i \leqslant T$ form a *closed class*. This chain is therefore not irreducible.   □

(b) *Random walks.* For unbounded random walks, Example (3.1.*d*), all states communicate and the chain is irreducible. For random walks with an absorbing boundary on one side the state 0 forms a *closed class* by itself. The state 0 is *absorbing*. The states $i > 0$ also form a class which is, however, *not* closed, for all these states lead to 0. For the random walks with two absorbing boundaries the boundaries 0 and k are absorbing states, while the states $0 < i < k$ again form a class which is not closed.   □

(c) *Branching processes.* In Example (3.1.*e*) the state 0 is absorbing. Whether the other states communicate depends on the distribution of the offspring Z. If, for example, the number of offspring can only be *even*, then all even states communicate whereas odd states cannot lead to themselves.   □

(d) *Stock-keeping.* If the demand distribution in Example (3.1.*f*) has not been chosen especially – like the distribution of the offspring in Example (*c*) – and the state space is restricted to $I = \{s + 1, s + 2, \ldots, S\}$, the chains are *irreducible*.   □

To every state i of a Markov chain one can associate a discrete *renewal process* by considering every return to the state i as a renewal. Every time the Markov chain enters the state i, it starts anew, *independently* of the past, and it is obvious that the consecutive moments $S_1 < S_2 < S_3 \ldots$, at which the Markov chain enters the state i, form a renewal process. This observation is extremely important since it allows us to extend the renewal theorems to Markov chains. The *renewal distribution* is given by $f_i^{(k)}, k = 1, 2, \ldots$, the probability that, starting from i, the next time i will be reached is after k steps. Let furthermore $f_{ji}^{(k)}, k = 1, 2, \ldots$, be the probability that, starting from j, the state i is reached for the *first time* after k steps (where $f_{ii}^{(k)} = f_i^{(k)}$) and let

$$f_i = \sum_{k=1}^{\infty} f_i^{(k)} \leqslant 1, \quad f_{ji} = \sum_{k=1}^{\infty} f_{ji}^{(k)} \leqslant 1, \quad f_{jj} = f_j.$$

f_i is the probability that the Markov chain starts from i and returns to i after finitely many steps and f_{ji} is the probability that the chain, when starting from j, will at some stage reach i.

According to renewal theory (especially Section 2.2) we call a state i *recurrent* if $f_i = 1$, and *transient* if $f_i < 1$. By the theory of discrete renewal processes (Section 2.2) it follows at once that a recurrent state i returns to itself *infinitely* often with probability 1, while a transient state leads to itself only a *finite* number of times. Similarly we accept the terminology of periodicity from renewal theory. A state i is called *periodic* with *period* $d > 1$ if d is the largest integer such that $f_i^{(k)} = 0$, if d is not a divisor of k.

If the Markov chain at time n is in state i, then either it is occurring for the *first time* or it has previously been in state i where the last time it occurred was at a time $n - r, r = 1, \ldots, n - 1$, and reaches the state i for the *next time* after r steps. If the chain at time $k = 0$ is in the state j, then the first event has probability $f_{ji}^{(n)}$ and the second set of events has the probabilities $p_{ji}^{(n-r)} f_i^{(r)}$. If we sum we obtain

$$p_{ji}^{(n)} = f_{ji}^{(n)} + \sum_{r=1}^{n-1} p_{ji}^{(n-r)} f_i^{(r)}, \quad n = 1, 2, \ldots. \tag{1}$$

This is a renewal equation of the form (2.1.14) for $p_{ij}^{(n)}$, where $p_{ji}^{(n)}$ corresponds to m_n in (2.1.14). The argument for the derivation of the renewal equation (1) is, of course, just a repetition of the derivation of (2.1.14). If the initial state is $j = i$, it follows from (1) that

$$p_{ii}^{(n)} = f_i^{(n)} + \sum_{r=1}^{n-1} p_{ii}^{(n-r)} f_i^{(r)}, \quad n = 1, 2, \ldots. \tag{2}$$

For reasons of completeness let us mention a second way in which the $p_{ji}^{(n)}$ can be determined. The chain is in the state i for the *first time* at time n or at time $n - r, r = 1, 2, \ldots, n - 1$, where it returns to i in the remaining r steps. From this it follows that

$$p_{ji}^{(n)} = f_{ji}^{(n)} + \sum_{r=1}^{n-1} f_{ji}^{(n-r)} p_{ii}^{(r)}, \quad n = 1, 2, \ldots. \tag{3}$$

For $j = i$ (3) again becomes (2).

Theorem 2.2.1 yields a useful criterion for recurrence and transience for the states of a Markov chain.

Theorem 2

(a) A state i is *recurrent* if and only if

$$\sum_{n=1}^{\infty} p_{ii}^{(n)} = \infty. \tag{4}$$

(b) A state i is *transient* if and only if

$$\sum_{n=1}^{\infty} p_{ii}^{(n)} < \infty. \tag{5}$$

(c) The series

$$\sum_{n=1}^{\infty} p_{ji}^{(n)} \tag{6}$$

converges if i is *transient*. If i is *recurrent* and $j \Rightarrow i$, the series *diverges*; if j does not lead to i, all the terms in (6) are equal to zero.

Proof. Theorem 2 is an immediate consequence of Theorem 2.2.1 since the $p_{ii}^{(n)}$ and $p_{ji}^{(n)}$ correspond as solutions of (2) and (1), respectively, to the m_n as solutions of (2.1.14). In (c) we just have to consider the trivial special case not considered in Theorem 2.2.1, that it may happen that $f_{ji} = 0$ (if j does not lead to i). ∎

Before we determine the asymptotic behaviour of the Markov chain's transition probabilities with the help of the renewal theorems, we have to comment on the subject of periodicity. It is obvious that $p_{ii}^{(n)} = 0$ if the period d is not a divisor of n, since renewals can only take place at a multiple of d. We have

Theorem 3
All states of a *class* have *equal* period d.

Proof. Let i and j be states of the same class, so that $i \Leftrightarrow j$, and let the periods of i and j be d_i and d_j. Then there exist integers $m > 0$ and $n > 0$ such that $p_{ij}^{(m)} > 0$ and $p_{ji}^{(n)} > 0$. If $p_{ii}^{(s)} > 0$, it follows that $p_{jj}^{(n+s+m)} \geq p_{ji}^{(n)} p_{ii}^{(s)} p_{ij}^{(m)} > 0$; similarly $p_{ii}^{(2s)} \geq p_{ii}^{(s)} p_{ii}^{(s)} > 0$ implies that $p_{jj}^{(n+2s+m)} > 0$. Therefore d_j is a divisor of $n + 2s + m - (n + s + m) = s$. Since all s with $p_{ii}^{(s)} > 0$ have d_i as their highest common factor, d_j has to be a divisor of d_i. If we interchange the roles of i and j in this argument, it follows that d_i is also a divisor of d_j, and thus $d_i = d_j$. ∎

A class whose states have period $d = 1$ is called *aperiodic*. In particular we call an irreducible chain aperiodic if the state space is an aperiodic class.
Let

$$\mu_i = \sum_{n=1}^{\infty} n f_i^{(n)} \tag{7}$$

be the expected value of the renewal distribution for the state i. We obtain from Theorem 2 and the renewal theorem (Theorem 2.2.2), applied to the renewal equations (1), (2):

Theorem 4

For each state i exactly one of the following statements holds:

(a) i is *transient*. In this case $\lim_{n\to\infty} p_{ji}^{(n)} = 0$ for all j.

(b) i is *recurrent* and $\mu_i = \infty$. In this case $\lim_{n\to\infty} p_{ji}^{(n)} = 0$ for all j; i is called *null recurrent*.

(c) i is *recurrent, aperiodic* and $\mu_i < \infty$. In this case $\lim_{n\to\infty} p_{ji}^{(n)} = f_{ji}/\mu_i$ for all j; i is called *positively recurrent*.

(d) i is recurrent, periodic with period d and $\mu_i < \infty$. In this case $\lim_{n\to\infty} p_{ii}^{(nd)} = d/\mu_i$; i is also called *positively recurrent*.

Proof. (a) follows from Theorem 2. (b) and (c) are direct consequences of the renewal theorem (Theorem 2.2.2), where f_{ji} takes the role of the g occurring there. Note that in (c) it may be that $f_{ji} = 0$ (if j does not lead to i). (d) follows from the corollary to Theorem 2.2.2. ∎

Like periodicity, transience and recurrence are properties which are shared by all states of a class. This shows clearly that the states of one class form a family whose members are closely linked by similar properties.

Theorem 5

For every *class* exactly one of the following statements holds:

(a) All states of the class are *transient*.

(b) All states of the class are *null recurrent*.

(c) All states of the class are *positively recurrent*.

Proof. Let the set C of states be a class and let $j, k \in C$. Then there exist integers $n > 0$ and $m > 0$ such that $\alpha = p_{jk}^{(n)} > 0$ and $\beta = p_{kj}^{(m)} > 0$, since $j \Rightarrow k$ and $k \Rightarrow j$. For $s > 0$ it follows that

$$p_{jj}^{(s+n+m)} \geqslant p_{jk}^{(n)} p_{kk}^{(s)} p_{kj}^{(m)} = \alpha\beta p_{kk}^{(s)}, \tag{8}$$

$$p_{kk}^{(s+n+m)} \geqslant p_{kj}^{(m)} p_{jj}^{(s)} p_{jk}^{(n)} = \alpha\beta p_{jj}^{(s)}. \tag{9}$$

If j is recurrent (transient) we have by Theorem 2

$$\sum_{s=1}^{\infty} p_{jj}^{(s)} = \infty \, (<\infty).$$

From (9) and (8) we conclude that

$$\sum_{s=1}^{\infty} p_{kk}^{(s)} = \infty \, (<\infty).$$

According to Theorem 2 the state k is also recurrent (transient). If j is null recurrent, i.e. if we have $p_{jj}^{(s)} \to 0$ for $s \to \infty$, it again follows from (8) that $p_{kk}^{(s)} \to 0$ and k is also null recurrent. ∎

According to Theorem 5 it is permitted to speak simply of (null or positively) recurrent or transient classes. From this theorem we conclude in particular that the states of an irreducible Markov chain are either all transient or all recurrent. In the latter case all states are either null recurrent or positively recurrent. In addition all states of an irreducible chain have the same period, by Theorem 3.

Theorem 6
If j is *recurrent* and if $j \Rightarrow i$, we also have $i \Rightarrow j$, i.e. $j \Leftrightarrow i$. Every *recurrent* class is *closed*.

Proof. If $j \Rightarrow i$, there exists an $n > 0$ such that $p_{ji}^{(n)} > 0$. If we did not have $i \Rightarrow j$, then the probability of no return from j to j would be at least equal to $p_{ji}^{(n)}$, hence $f_j \leqslant 1 - p_{ji}^{(n)} < 1$, and j would not be recurrent, contrary to our assumption. This shows that every class of recurrent states is closed. ∎

A transient class may, but need not necessarily, be closed. The following Theorem 7 shows that a *finite* class of transient states cannot be closed. However, there exist, for instance, irreducible Markov chains with countably *infinitely* many transient states (see the examples below). In this case the state space forms a closed transient class.

Theorem 7
(a) For a Markov chain with *finitely* many states there can be no null recurrent states and not all states can be transient.
(b) A class with *finitely* many transient states cannot be closed.

Proof. If there existed a finite class C of null recurrent states it would have to be closed by Theorem 6, and by Theorem 1 its states would define a stochastic transition matrix. Furthermore, $p_{jk}^{(n)} \to 0$ for $n \to \infty$ and all $j, k \in C$. At the same time the *finite* row sums of $p_{jk}^{(n)}$ over k for all n must be equal to 1. This is, however, impossible. If furthermore all states were transient they would form a finite and closed set. The same argument leads to a contradiction. (b) is proved in exactly the same way. ∎

Theorem 8
A class of states is *recurrent* if and only if, for every pair of states i, j of the class, we have $f_{ji} = 1$. In particular in a recurrent class every state is reached from every other state with probability 1.

Proof. Let the class be recurrent and let i, j be two states of this class. Then there exists an $n > 0$ such that $p_{ij}^{(n)} > 0$. The probability of no return from i to i is equal to $1 - f_i$, and this probability is greater than

$p_{ij}^{(n)}(1 - f_{ji})$, since $1 - f_{ji}$ is the probability of no return from j to i. If $f_{ji} < 1$, then $f_i < 1$, and i would, contrary to the conditions, not be recurrent.

If, in contrast, $f_{ji} = 1$ for all j and i of this class, we have in particular $f_{jj} = f_j = 1$, and the class is recurrent. ∎

Let $\{p_i, i \in I\}$ be a *distribution* with the property that

$$p_i = \sum_j p_j p_{ji}, \quad i \in I. \tag{10}$$

If such a distribution is an initial distribution for the Markov chain, it follows from (3.1.12) that $p_i = p_i^{(1)} = p_i^{(2)} = \ldots = p_i^{(n)} = \ldots$. The absolute distributions of the states always remain the same. For this reason we call a distribution which satisfies (10) a *stationary distribution*. Stationary distributions can be defined, if they exist, as solutions of the linear equations (10) with the additional conditions $p_1 + p_2 + \ldots = 1$. However, there does not always exist a stationary distribution, and even if there exists one, it is not necessarily uniquely defined. This is shown by the next proposition.

Theorem 9

For an *irreducible aperiodic* Markov chain exactly one of the following two assertions holds:

(a) All states are *transient* or *null recurrent*. In this case there is *no* stationary distribution.

(b) All states are *positively recurrent*. In this case we have $f_{ji} = 1$ for all i, j and

$$u_i = \lim_{n \to \infty} p_{ji}^{(n)} = \frac{1}{\mu_i} > 0 \tag{11}$$

is the *only* stationary distribution. For every initial distribution we have $\lim_{n \to \infty} p_i^{(n)} = u_i$.

Before going into the proof we should like to add a few remarks. If a Markov chain has *several* positively recurrent classes, then each of them is closed by Theorem 6 and thus forms by itself an irreducible chain which, according to this theorem, possesses a unique stationary distribution. By setting the probability of all states which do not belong to the class about to be considered equal to zero, the stationary distribution of the class becomes a stationary distribution for the entire original chain. Every convex linear combination of these stationary distributions again forms a stationary distribution for the chain. This shows that a chain can have several stationary distributions. On the other hand, by Theorem 7, all states have to be positively recurrent for an irreducible Markov chain with *finitely* many states. According to this theorem there always exists a unique stationary distribution in this case.

Proof of Theorem 9. First we shall prove (*b*). From Theorem 4(*c*) it follows that

$$1 = \sum_i p_{ji}^{(n)} \geqslant \sum_{i=1}^{N} p_{ji}^{(n)} \to \sum_{i=1}^{N} u_i \quad \text{as } n \to \infty,$$

hence

$$\sum_i u_i \leqslant 1.$$

Similarly it follows from (3.1.10) that

$$p_{ji}^{(n+1)} = \sum_k p_{jk}^{(n)} p_{ki} \geqslant \sum_{k=1}^{N} p_{jk}^{(n)} p_{ki}.$$

Letting $n \to \infty$ and applying Theorem 4(*c*) we obtain

$$u_i \geqslant \sum_k u_k p_{ki}. \tag{12}$$

Summing this relation over i, we realize that in (12) there must be equality for all i. If we now multiply by p_{ij} and sum over i, it follows that (12) holds with $p_{ki}^{(2)}$ instead of p_{ki}; from this it follows that in general

$$u_i = \sum_k u_k p_{ki}^{(n)}, \quad n = 1, 2, \ldots . \tag{13}$$

Since $\sum u_i$ converges and $p_{ki}^{(n)}$ is uniformly bounded in n, we obtain for $n \to \infty$

$$u_i = \sum_k u_k u_i, \quad \text{for all } i.$$

This in turn implies

$$\sum_k u_k = 1. \tag{14}$$

Since in addition, according to Theorem 4(*c*), $u_i > 0$ for all i, it follows from (13) and (14) that $\{u_k, k \in I\}$ is a *stationary distribution* of the Markov chain. If $\{p_k, k \in I\}$ is an arbitrary initial distribution, it follows from (3.1.11) that

$$\lim_{n \to \infty} p_i^{(n)} = \sum_k p_k u_i = u_i.$$

If $\{v_k, k \in I\}$ is a stationary distribution, (13) holds for the v_k, and for $n \to \infty$ we obtain as before

$$v_i = \sum_k v_k u_i$$

and thus $v_i = u_i$. We have thus proved (*b*).

If $\{v_i, i \in I\}$ is again a stationary distribution but if we now have $p_{ki}^{(n)} \to 0$, it would follow from (13) that $v_i = 0$ for all i. This is a contradiction. Therefore there can be no stationary distribution in this case. We have thus also proved (a). ∎

One of the most important aspects of this theorem lies in the fact that it gives a useful criterion for deciding whether states of an irreducible, aperiodic chain are positively recurrent. If we succeed in defining a stationary distribution of a chain by solving (10), then the states must be positively recurrent, according to Theorem 9, since case (a) is excluded by this assumption.

Examples

(e) *Asymptotic distribution and lim P^n.* The chain with two states in Example (3.1.g) is irreducible. Thus, according to Theorem 7, both states are *positively recurrent*. The system of equations (10) for the stationary distribution in this example is

$$p_1 = \tfrac{1}{2}p_1 + \tfrac{1}{4}p_2, \quad p_2 = \tfrac{1}{2}p_1 + \tfrac{3}{4}p_2.$$

The only solution with $p_1 + p_2 = 1$ is $p_1 = \tfrac{1}{3}, p_2 = \tfrac{2}{3}$. This is in accordance with Theorem 9. These values can also be found in both rows of the limit matrix $\lim_{n \to \infty} P^n$ (see Example (3.1.g)). This is true in general for irreducible, positively recurrent chains. The jth row of P^n contains the values $p_{ji}^{(n)}$, $i = 1, 2, \ldots$, and these tend to $u_i = 1, 2, \ldots$, independently of j. From Theorem 9 we also conclude the mean renewal times $\mu_1 = 1/p_1 = 3$, $\mu_2 = 1/p_2 = \tfrac{3}{2}$ for the states 1 and 2. ☐

(f) *Discrete renewal processes.* Let us first consider the chain which has been associated with the age of the renewal process at time n (see Example (3.1.b)). The equations (10) in this example are

$$p_1 = \frac{\sum_{i=1}^{\infty} p_i f_i}{1 - \sum_{k=1}^{i-1} f_k},$$

$$p_i = \frac{p_{i-1}(1 - \sum_{k=1}^{i-1} f_k)}{1 - \sum_{k=1}^{i-2} f_k}, \quad i > 1.$$

It is simple to express $p_2, p_3, \ldots$ in terms of p_1 by substitution. We find that

$$p_i = \left(1 - \sum_{k=1}^{i-1} f_k\right) p_1, \quad i > 1.$$

If the renewal process is aperiodic, then the same is true for the chain considered here, for, by the definition of an aperiodic renewal process (Section 2.2), the state 1 is aperiodic and so are all the others by Theorem 3, since the chain is irreducible (see Example (a)). By Theorem 9 the states are positively recurrent if there exists a stationary distribution and this is the case if

$$\sum_{i=1}^{\infty} \left(1 - \sum_{k=1}^{i-1} f_k\right) > 0,$$

If in this case we write

$$p_1 = \frac{1}{\sum_{i=1}^{\infty} (1 - \sum_{k=1}^{i-1} f_k)} > 0,$$

then all p_i are *positive* and the sum of the p_i is equal to 1.

If the renewal process is *transient*, i.e. $f_1 + f_2 + \ldots < 1$, the condition is certainly not satisfied. If, however, the renewal process is *recurrent*, i.e. $f_1 + f_2 + \ldots = 1$, we have

$$\sum_{i=1}^{\infty} \left(1 - \sum_{k=1}^{i-1} f_k\right) = \sum_{i=1}^{\infty} \sum_{k=i}^{\infty} f_k = \sum_{n=1}^{\infty} n f_n = \mu,$$

where μ is the mean renewal time of the renewal process. The *Markov chain* is thus *positively recurrent*, if the *renewal process is recurrent* with $\mu < \infty$. The stationary distribution is

$$p_1 = \frac{1}{\mu}; \quad p_i = \frac{(1 - \sum_{k=1}^{i-1} f_k)}{\mu}, \quad i > 1.$$

This agrees with the corresponding result in Example (2.2.c).

For the chain connected with the remaining life the equations (10) (see Example (3.1.b)) are given by

$$p_i = p_{i+1} + p_i f_i, \quad i \geq 1.$$

We can easily verify that

$$p_i = \left(1 - \sum_{k=1}^{i-1} f_k\right) p_1, \quad i > 1,$$

so that any further investigation proceeds just as above and leads to the same results, see also Example (2.2.c). $\square$

The existence of a stationary distribution as solution of the equations (10) can, as in Example (*f*), also be proved in many other cases by explicit determination of the solution. As already mentioned, Theorem 9 becomes a very useful criterion for proving the positive recurrence of an aperiodic class or an irreducible aperiodic Markov chain. If, however, no stationary distribution exists, the question remains whether the states are transient or null recurrent. A useful criterion for this is given by the following theorem.

Theorem 10

The states of an *irreducible* Markov chain are *transient* if and only if the system of equations

$$y_i = \sum_{j \neq k} p_{ij} y_j, \quad i \in I, \text{but } i \neq k \tag{15}$$

has a bounded non-zero solution, where $k \in I$ can be chosen arbitrarily.

Proof. Let $y_i^{(n)}$ be the probability that the chain, starting from i, has *not* reached the state k up to and including time n. We have

$$y_i^{(1)} = \sum_{j \neq k} p_{ij}, \quad y_i^{(n+1)} = \sum_{j \neq k} p_{ij} y_j^{(n)}. \tag{16}$$

It follows from this that $y_i^{(1)} \leqslant 1$ and thus $y_i^{(2)} \leqslant y_i^{(1)}$ and, by induction on n, $y_i^{(n+1)} \leqslant y_i^{(n)}$. For if $y_i^{(n)} \leqslant y_i^{(n-1)}$, then

$$y_i^{(n+1)} = \sum_{j \neq k} p_{ij} y_j^{(n)} \leqslant \sum_{j \neq k} p_{ij} y_j^{(n-1)} = y_i^{(n)}.$$

Therefore $y_i = \lim_{n \to \infty} y_i^{(n)}$ exists. y_i is the probability that, starting from i, the state k is *never* reached, i.e. $y_i = 1 - f_{ik}$.

According to Theorem 8, $f_{ik} = 1$ and therefore $y_i = 0$ for all $i \neq k$, if the chain is recurrent. Consequently, the Markov chain, starting from k, returns to k at a later time if it returns to k with its next step or if it changes over to the state $i \neq k$ with its next step and then, starting from i, again returns to k. We have therefore

$$f_k = p_{kk} + \sum_{i \neq k} p_{ki} f_{ik}.$$

If we now have $y_i = 0$ for all $i \neq k$ and therefore $f_{ik} = 1$, then $f_k = 1$ and thus k and, by Theorem 5, the whole chain are recurrent. The chain is therefore transient if and only if for at least one $i \neq k$, $y_i > 0$.

If we let $n \to \infty$ in (16) we realize that the y_i satisfy the equations (15). If z_i is an arbitrary *bounded non-zero* solution of (15), we can normalize z_i in such a way that $|z_i| \leqslant 1$. By comparing (15) – written in terms of z_i – with (16) we see that $|z_i| \leqslant y_i^{(1)}$, since

$$|z_i| \leqslant \sum_{j \neq k} p_{ij} |z_j| \leqslant \sum_{j \neq k} p_{ij} = y_i^{(1)}.$$

From this it follows furthermore by induction on n that $|z_i| \leqslant y_i^{(n)}$, since $|z_i| \leqslant y_i^{(n-1)}$ implies

$$|z_i| \leqslant \sum_{j \neq k} p_{ij} |z_j| \leqslant \sum_{j \neq k} p_{ij} y_j^{(n-1)} = y_i^{(n)}.$$

Therefore we also have $|z_i| \leqslant y_i$. Thus y_i is the *maximal* solution of (15) bounded by 1. It follows that $y_i = 0$ for all i if and only if (15) has no non-zero bounded solution. This proves the theorem. ∎

Examples

(g) *Discrete renewal processes.* We have shown in Example (f) that
the states of a Markov process associated with age are positively recurrent
only if the renewal process itself is recurrent and the mean renewal time μ
is *finite*. If, however, $\mu = \infty$, are the states null-recurrent or transient? The
reader can deduce from renewal arguments that the states are in this case
recurrent or transient according as to whether the renewal process is recurrent
or transient. This result will now be deduced from Theorem 10. In this
example the system (15) for $k = 1$ becomes (see Example (3.1.b)):

$$y_i = \frac{y_{i+1}(1 - \Sigma_{k=1}^{i} f_k)}{1 - \Sigma_{k=1}^{i-1} f_k}, \quad i > 1.$$

By expressing $y_3, y_4, \ldots$ in turn in terms of y_2 we find

$$y_i = \frac{y_2(1 - f_1)}{1 - \Sigma_{k=1}^{i-1} f_k}, \quad i \geqslant 2.$$

If $f = f_1 + f_2 + \ldots = 1$, i.e. if the renewal process is recurrent, there does
not exist a non-zero bounded solution. The states have thus to be recurrent
by Theorem 10 and in particular null-recurrent for $\mu = \infty$. If the renewal
process is transient, then all states of the chain are transient, since for $f < 1$
the system (15) has non-zero bounded solutions. $\qquad\square$

(h) *Random walks.* We first note that the random walks introduced in
Example (3.1.d) have states with period $d = 2$ apart from the absorbing
states. Theorem 9 can therefore not be applied directly to random walks. We
can, however, apply Theorem 10. For an unbounded random walk we may
choose $k = 0$, for which the system (15) will read as follows, if we put $y_0 = 0$:

$$y_i = py_{i-1} + qy_{i+1}, \quad i = \pm 1, \pm 2, \ldots, \quad y_0 = 0.$$

If $p = q = \frac{1}{2}$, it follows recursively that $y_i = iy_1, i \geqslant 1$, and $y_i = -iy_{-1}$,
$i \leqslant -1$, and that there exists no non-zero bounded solution. The states of an
unbounded random walk in this case are *recurrent*, according to Theorem 10.
If $p \neq q$ but $p + q = 1$ it follows in general that $(p + q)y_i = py_{i-1} + qy_{i+1}$ or
$y_{i+1} - y_i = (p/q)(y_i - y_{i-1})$ and hence that $y_{i+1} - y_i = (p/q)^i y_1$ for $i \geqslant 0$.
By summation of this relation over $i = 1, \ldots, j - 1$ we obtain

$$y_j = \frac{y_1(1 - (p/q)^j)}{1 - p/q}, \quad j \geqslant 1.$$

Similarly we obtain for $j = -1, -2, \ldots$

$$y_j = \frac{y_{-1}(1 - (q/p)^{-j})}{1 - q/p}, \quad j \leqslant -1.$$

If $p < q$, we obtain bounded non-zero solutions, if we choose $y_1 \neq 0$ and

$y_{-1} = 0$. The states in this case are *transient* by Theorem 10. The maximal solution of (15), bounded by 1, is given by $y_j = 0$ for $j \leqslant -1$ and

$$y_j = (1 - (p/q)^j), \quad j \geqslant 1.$$

According to the proof of Theorem 10 these y_j represent the probabilities that, starting from j, the state 0 is *never* reached. In the case of a random walk with one-sided absorbing boundary with $p < q$ the probability of an absorption in 0, starting from $j \leqslant -1$, is equal to 1 and for $j > 1$ this absorption probability is equal to $(p/q)^j$. The case $q < p$ is symmetric to the case just considered. In particular the unbounded random walk is *transient* if $p \neq q$. $\qquad\qquad\qquad\qquad\qquad\qquad\qquad\qquad\qquad\qquad\qquad\qquad$ $\square$

3.3. Markov chains with continuous time parameter

Let $\{X_t, 0 \leqslant t < \infty\}$ be a family of random variables with a *countable* range I. The index t can normally be interpreted as *time*, and $\{X_t, 0 \leqslant t < \infty\}$ is called a *stochastic process with continuous time parameter*, or simply *continuous parameter*. Since the range of the X_t is countable for all t, all variables X_t are discrete (see Section 1.2). As already mentioned in Section 3.1, I is called the *state space* and the elements of I are called *states*. If X_t takes the value $i \in I$, we speak of the process as being in state i at time t.

Let $0 \leqslant t_1 < \ldots < t_n$ be an arbitrary n-tuple of times and $i_1, \ldots, i_n$ an associated n-tuple of states in $I, n \geqslant 2$. If we always have

$$P[X_{t_n} = i_n | X_{t_1} = i_1, \ldots, X_{t_{n-1}} = i_{n-1}] = P[X_{t_n} = i_n | X_{t_{n-1}} = i_{n-1}], \quad (1)$$

the process $\{X_t, 0 \leqslant t < \infty\}$ is called a *Markov chain with continuous parameter*. Condition (1) has exactly the same form as (3.1.1), except that the t are here real numbers instead of integers. We can therefore expect that the theories of the Markov chains with continuous and discrete parameters will be analogous. Condition (1) is again called the *Markov property*.

Exactly as in Section 3.1 (see Theorem 3.1.1) we can show that

$$P[X_{t_v} = i_v; n \leqslant v \leqslant n + m | X_{t_v} = i_v; 1 \leqslant v \leqslant n - 1]$$

$$= P[X_{t_v} = i_v; n \leqslant v \leqslant n + m | X_{t_{n-1}} = i_{n-1}] \quad (2)$$

if $0 \leqslant t_1 < \ldots < t_n < \ldots < t_{n+m}, i_v \in I$ for $v = 1, \ldots, n + m, n \geqslant 2, m \geqslant 0$. In the continuous case the probabilities of future states thus depend again only on the last *observed* state and not on the previously observed states.

Condition (3.1.3) also holds for Markov chains with continuous parameter, where t can, of course, take real values. The probabilities $P[X_{t''} = j | X_{t'} = i]$, $t' < t''$, which appear there are called *transition probabilities*. They describe the probability that the process will be in the state j at time t'' if it is in state i at time t'. These transition probabilities are called *stationary* if $P[X_{t''} = j | X_{t'} = i] = P[X_{t''+h} = j | X_{t'+h} = i]$ for all $t' < t''$ and h. In this case the transition probabilities do not depend on t', t'' but only on the differences $s = t'' - t'$. We can, therefore, define the *transition functions* $p_{ij}(s) = P[X_{t+s} = j | X_t = i]$. In the following *we shall only consider Markov chains with stationary probabilities.*

For every fixed s and for every fixed initial state i, $\{p_{ij}(s), j \in I\}$ is a discrete distribution so that we must have

$$p_{ij}(s) \geqslant 0 \text{ for all } j \in I, \ \sum_j p_{ij}(s) = 1. \tag{3}$$

All summations are taken over the state space I unless explicitly stated otherwise. If $0 \leqslant t_0 < t_1 < \ldots < t_n$, it follows (e.g. from (3.1.3)) that

$$P[X_{t_\nu} = i_\nu; 1 \leqslant \nu \leqslant n | X_{t_0} = i_0] = \prod_{\nu=1}^{n} p_{i_{\nu-1} i_\nu}(t_\nu - t_{\nu-1}), \tag{4}$$

and in particular that

$$P[X_{t_0+s} = k, X_{t_0+s+t} = j | X_{t_0} = i] = p_{ik}(s) p_{kj}(t).$$

If we sum this relation over $k \in I$ we obtain on the left hand side $P[X_{t_0+s+t} = j | X_{t_0} = i] = p_{ij}(s + t)$ and thus

$$p_{ij}(s + t) = \sum_k p_{ik}(s) p_{kj}(t), \tag{5}$$

the *Chapman–Kolmogorov equations* in the continuous case (see Theorem 3.1.2).

As in the case of a discrete parameter we can introduce here the *absolute distributions* $p_i(t) = P[X_t = i]$. The distribution $p_i = p_i(0)$ is the *initial distribution* of the Markov chain. We have

$$p_i(t) = P[X_t = i] = \sum_k P[X_t = i | X_0 = k] \cdot P[X_0 = k] = \sum_k p_k p_{ki}(t), \tag{6}$$

(see (1.1.17)). It thus follows for the finite-dimensional distributions at times $0 \leqslant t_0 < t_1 \ldots < t_n$ from (3.1.3) that

$$P[X_{t_0} = i_0, X_{t_1} = i_1, \ldots, X_{t_n} = i_n] = \sum_k p_k p_{ki_0}(t_0) \prod_{\nu=1}^{n} p_{i_{\nu-1} i_\nu}(t_\nu - t_{\nu-1}). \tag{7}$$

Equation (7) shows that a Markov chain with continuous parameter and stationary transition probabilities is defined by the initial distribution p_i and the transition probability functions $p_{ij}(s)$ which have to satisfy (3) and (5).

The matrix notation enables us to summarize some formulae very clearly. Let $P(s)$ be the matrix $(p_{ij}(s))$ of the transition probabilities and $p(t)$ the vector of the absolute probabilities, $p = p(0)$. The Chapman–Kolmogorov equations (5) are now $P(s + t) = P(s)P(t)$, and (6) reads as $p(t) = pP(t)$.

In spite of all the similarity between the theories of Markov chains with discrete and continuous parameters we have to note a few important differences: for the discrete parameter the *one-step* transition probabilities p_{ij} generate, so to speak, the Markov chain. In particular the *multiple-step* transition probabilities can be derived from them (see (3.1.10)). For a continuous parameter we do not yet have an analogue to the 'one-step' transition probabilities; the $p_{ij}(s)$ are 's-step' transition probabilities, so to speak. $p_{ij}(1)$ cannot play the same role as the one-step probabilities for discrete Markov chains. We have to find 'generating elements' out of which the $p_{ij}(s)$ can be constructed, just as $p_{ij}^{(n)}$ can be constructed out of p_{ij}. This question is of practical importance since it would be very inconvenient if all the transition probability functions $p_{ij}(s)$ had to be given as data, if a Markov chain with continuous parameter is to be used as a model of a real process.

In addition to conditions (3) and (5) we shall now introduce rather strong *regularity conditions* for the transition probability functions. Essentially, we are concerned with those conditions which one assumes initially when applying Markov chains with continuous parameter as models for real processes.

The first condition is that the probability of a *change of state* during a time interval of length t should for short intervals, be roughly proportional to t. This can be ensured by the condition $1 - p_{ii}(t) = q_i t + o_i(t)$ where in general $o(t)$ is a quantity which tends to 0 faster than t, i.e. $(o(t)/t) \to 0$ as $t \to 0$. This condition is equivalent to

$$\lim_{t \to 0} \frac{1 - p_{ii}(t)}{t} = q_i, \quad 0 \leqslant q_i < \infty, \text{ for all } i \in I. \tag{8}$$

If during the short time interval t a change of state does take place, the probability for the transition from the initial state i to the new state j should be essentially independent of the length t of the interval if t is small. The *conditional* probability for a transition to j, given the occurrence of a transition is $p_{ij}(t)/(1 - p_{ii}(t))$ and this probability is to be equal to $q'_{ij} + o'_{ij}(t)$. Here the q'_{ij} have to form a stochastic matrix with $q'_{ii} = 0$. If we write $q_{ij} = q'_{ij} q_i$, we conclude from (8) that this condition is equivalent to

$$\lim_{t \to 0} \frac{p_{ij}(t)}{t} = q_{ij} \text{ for all } i,j \in I, i \neq j, \quad \sum_{j \neq i} q_{ij} = q_i, \ q_{ij} \geqslant 0. \tag{9}$$

This means that the $p_{ij}(t)$ are themselves of the form $q_{ij} t + o_{ij}(t)$.

Out of the elements q_i and q_{ij} we can thus form a matrix $A = (a_{ij})$ with $a_{ii} = -q_i$ and $a_{ij} = q_{ij}$ for $i \neq j$. We have $a_{ii} \leqslant 0$ and $a_{ij} \geqslant 0$, and the row sums of the matrix A are, by (9), equal to zero. Such a matrix is called a *differential matrix*.

Finally, we impose the following third additional condition:

For $t \to 0$, $p_{ij}(t)/t$ converges for all $j \in I$ uniformly in i to q_{ij}. \qquad (10)

For Markov chains with a finite state space this condition is, of course, automatically satisfied. It is thus only significant for infinite state spaces. Its significance will become clearer in the proof of Theorem 1 below.

A direct consequence of the regularity conditions (8), (9) is that we have

$$\lim_{t \to 0} p_{ii}(t) = 1, \quad \lim_{t \to 0} p_{ij}(t) = 0 \text{ for } i \neq j \tag{11}$$

and that the $p_{ij}(t)$ are differentiable at $t = 0$. The following theorem shows furthermore that under the conditions (8) to (10) the $p_{ij}(t)$ are differentiable for all t and that they satisfy two systems of linear differential equations in which the q_i and q_{ij} appear as coefficients.

Theorem 1

According to conditions (8), (9) and (10) the transition probability functions $p_{ij}(t)$ are differentiable for all $t \geqslant 0$ and satisfy the *differential* equations

$$\dot{p}_{ij} = -q_i p_{ij} + \sum_{k \neq i} q_{ik} p_{kj}, \tag{12}$$

$$\dot{p}_{ij} = -p_{ij} q_j + \sum_{k \neq j} p_{ik} q_{kj}. \tag{13}$$

Equations (12) are *Kolmogorov's backward equations* and (13) the *forward equations*.

Proof. It follows from the Chapman–Kolmogorov equations (5) that

$$p_{ij}(s + t) - p_{ij}(t) = \sum_{k \neq i} p_{ik}(s) p_{kj}(t) - (1 - p_{ii}(s)) p_{ij}(t).$$

If we divide this relation by s and use (8), (9), we obtain

$$\frac{p_{ij}(s + t) - p_{ij}(t)}{s} = -q_i p_{ij}(t) + \sum_{k \neq i} q_{ik} p_{kj}(t) + \frac{o(s)}{s}. \tag{14}$$

In (14) the term $o(s)/s$ needs special justification. It is produced by adding the terms $o_{ik}(s)/s$ which appear in $p_{ik}(s) = q_{ik}s + o_{ik}(s)$ and contains furthermore $-o_i(s)/s$ from $1 - p_{ii}(s) = sq_i + o_i(s)$. We have to show that

$$\sum_{k \neq i} \frac{o_{ik}(s)}{s} p_{kj}(t) \tag{15}$$

is of the form $o(s)/s$, i.e. that it again converges to 0 as $s \to 0$. The sum (15) can be decomposed into a partial sum over $k \leqslant N$ and the remaining series for $k > N$. For $N > i$ we have

$$\left| \sum_{k>N} \frac{o_{ik}(s)}{s} p_{kj}(t) \right| = \left| \sum_{k>N} \left(\frac{p_{ik}(s)}{s} - q_{ik} \right) p_{kj}(t) \right|$$

$$\leqslant \frac{1}{s} \sum_{k>N} p_{ik}(s) p_{kj}(t) + \sum_{k>N} q_{ik} p_{kj}(t)$$

$$\leqslant \frac{1}{s} \sum_{k>N} p_{ik}(s) + \sum_{k>N} q_{ik}$$

$$= \frac{1}{s} \left(1 - \sum_{k \leqslant N} p_{ik}(s) \right) + \left(q_i - \sum_{k \leqslant N} q_{ik} \right)$$

$$= \frac{1}{s} \left(1 - p_{ii}(s) - \sum_{\substack{k \leqslant N \\ k \neq i}} p_{ik}(s) \right) + \left(q_i - \sum_{k \leqslant N} q_{ik} \right)$$

$$\to 2 \left(q_i - \sum_{k \leqslant N} q_{ik} \right) \quad \text{as } s \to 0.$$

For sufficiently large N and sufficiently small s we can thus make the remaining series smaller than $\frac{1}{2}\epsilon$. The same is true for the partial sum over $k \leqslant N$. Thus (15) becomes smaller than ϵ and (14) is justified.

Thus the limit of the right hand side of (14) exists for $s \to 0$. Hence the limit of the left hand side in (14) exists and $\dot{p}_{ij}(t)$ exists for all $t \geqslant 0$, and (12) follows from (14). Note that in order to prove (12) condition (10) has not been used.

Similarly we obtain from (5), after dividing by t and using (8), (9),

$$\frac{p_{ij}(s+t) - p_{ij}(s)}{t} = -p_{ij}(s)q_j + \sum_{k \neq j} p_{ik}(s)q_{kj} + \frac{o(t)}{t}. \tag{16}$$

The term $o(t)/t$ in (16) is immediately justified from condition (10). Contrary to the backward equations the forward equations cannot be obtained without (10). Letting $t \to 0$ in (16) yields (13). ∎

If we differentiate (6) with respect to t and use the forward equations (13), we obtain from (6) as a corollary to Theorem 1 a system of differential equations for the absolute distributions $p_i(t)$.

Corollary
Under the conditions of Theorem 1 the $p_j(t)$ are differentiable for all $t \geqslant 0$ and we have

$$\dot{p}_j = -p_j q_j + \sum_{k \neq j} p_k q_{kj}. \tag{17}$$

Note that no backward equations can be obtained for absolute distributions. In matrix notation (12) and (13) are $\dot{P} = AP$ and $\dot{P} = PA$ where P is the matrix of the transition probability functions and A is the differential matrix introduced above.

On account of Theorem 1 we can consider the q_i and q_{ij} as generating elements or data for the definition of Markov chains. By solving (12) and (13) we are able to define from these the transition probability functions $p_{ij}(t)$. Indeed, continuous Markov chains as models for real processes are normally established by starting from conditions (8) and (9). In this connection we have to ask the following questions:

1. Do the equations (12), (13) have solutions $p_{ij}(t)$, which satisfy conditions (3) and (5) for the transition probability functions, for all q_i, q_{ij} which satisfy the second condition in (9)?
2. Is this solution unique?

For those cases which occur in applications one may assume that the answer to both questions is yes. One of the traps of the theory of Markov chains with continuous parameter is, however, that question 1 can be answered in the affirmative only with a modification, while question 2 must generally be answered in the negative. These phenomena only occur with an infinite state space. There may be solutions in which in the second condition in (3) we have $\leqslant 1$ instead of $= 1$ (substochastic transition probabilities). Such cases are to be ignored in the following considerations since we hardly ever come across them in practical applications. These questions will be pursued further in Section 6.3.

Examples

(*a*) *Poisson process.* We consider a system in which certain events can occur at arbitrary times. Let the probability that an event occurs during the time interval $[t, t + h)$ be equal to $\lambda h + o(h)$, *independently* of which events have occurred during the interval $[0, t)$. Let the probability that two or more events occur during the interval $[t, t + h)$ be $o(h)$. Let X_t be the number of events which have occurred during the interval $[0, t)$. More generally we allow X_0 to be equal to $i \geqslant 0$ so that, up to time 0, i events have already occurred. Similarly X_t is the number of events up to t, including those events which have occurred up to $t = 0$. $\{X_t, t \geqslant 0\}$ is a *continuous Markov chain* since the number of events during an interval $[t, t + s)$ is obviously independent of the number of events during the interval $[0, t)$. Let us denote its transition probability functions by $p_{ij}(t)$. We have $p_{ij}(t) = 0$ for $j < i$ and therefore $q_{ij} = 0$ for $j < i$, since jumps are only possible from i to $j > i$. Also, $p_{ii+1}(t)$ is equal to the probability of one event during the interval $[0, t)$. Therefore according to our initial assumption we have $p_{ii+1}(t) = \lambda t + o(t)$. Similarly it follows that $p_{ij}(t) = o(t)$ for $j > i + 1$. Thus we have (see (9)) $q_{ii+1} = \lambda, q_{ij} = 0$ for $j > i + 1$. Finally the probability of no event in the interval $[0, t)$ is equal to $1 - \lambda t + o(t)$ and thus $p_{ii}(t) = 1 - \lambda t + o(t)$ or $1 - p_{ii}(t) = \lambda t + o(t)$ which implies $q_i = \lambda$ (see (8)). Thus we have defined the differential matrix of the process.

The *forward equations* (13) for $p_{0j}(t) = p_j(t)$ for this process are

$$\dot{p}_0 = -\lambda p_0,$$
$$\dot{p}_j = -\lambda p_j + \lambda p_{j-1}, \quad j > 0.$$

The solution of these equations with the initial conditions $p_0(0) = 1$, $p_j(0) = 0$ for $j > 0$ is

$$p_j(t) = e^{-\lambda t}\frac{(\lambda t)^j}{j!}, \quad j \geqslant 0.$$

A comparison with Example $(2.1.a)$ shows that the process considered must obviously correspond to the *Poisson process* which thus does not just appear as a renewal process but also as a Markov chain.   □

(b) *Birth and death process.* Markov chains in which the only jumps are from a state i to an *adjacent* state $i + 1$ (birth) or $i - 1$ (death) are called birth and death processes. With these processes we assume $q_{ij} = 0$ for $j \neq i - 1$ and $j \neq i + 1$. In addition we make the stronger assumption that the probability of two or more jumps in the interval $[t, t + h)$ is of the order $o(h)$. This guarantees condition (10). If we put $q_{ii+1} = \lambda_i$ and $q_{ii-1} = \mu_i$ ($\mu_0 = 0$) we must have $q_i = \lambda_i + \mu_i$, and we obtain $p_{ii+1}(t) = \lambda_i t + o(t), p_{ii-1}(t) = \mu_i t + o(t)$, $p_{ij}(t) = o(t)$ for $j \neq i - 1$ and $j \neq i + 1$ and $1 - p_{ii}(t) = (\lambda_i + \mu_i)t + o(t)$. The backward equations (12) for birth and death processes are

$$\dot{p}_{0j} = -\lambda_0 p_{0j} + \lambda_0 p_{1j},$$
$$\dot{p}_{ij} = -(\lambda_i + \mu_i)p_{ij} + \lambda_i p_{i+1j} + \mu_i p_{i-1j}, \quad i > 0,$$

and the forward equations (13) are

$$\dot{p}_{i0} = -p_{i0}\lambda_0 + p_{i1}\mu_1,$$
$$\dot{p}_{ij} = -p_{ij}(\lambda_j + \mu_j) + p_{ij-1}\lambda_{j-1} + p_{ij+1}\mu_{j+1}, \quad j > 0.$$

If all $\mu_i = 0$ one speaks of a pure *birth process*. The Poisson process, Example (a), is a birth process with $\lambda_i = \lambda$ for all i.   □

(c) *Queueing systems.* Birth and death processes occur especially in simple queueing systems. Such systems will be considered in greater detail in Chapter 4. Here we shall only discuss a simple special case. Suppose m tills are open to serve customers. If a customer arrives and finds a server who is not busy, he is served at once. If all servers are busy serving customers who arrived previously, the newly arrived customer has to wait in line. A customer who has been served disappears from the system and makes room at the till for the next customer.

For the arrival process and serving procedure we make the following special assumptions: let the probability of an arrival during interval $[t, t + h)$ be equal to $\lambda h + o(h)$, independently of all previous arrivals, and let the probability of two or more arrivals in the same interval be $o(h)$. A comparison with Example (a) shows that the arrivals form a *Poisson process*. If a server is busy at time t let the probability that the server will complete the service in the interval $[t, t + h)$ be equal to $\mu h + o(h)$, independently of how long the service has already lasted. The probability that the server can complete two or more services in this interval is assumed to be equal to $o(h)$. Let these events for the various servers be mutually independent of the arrival times.

If at time t altogether s servers are busy, $0 < s \leqslant m$, then the probability that during the interval $[t, t + h)$ *exactly one* service is completed, is equal to

$$s(\mu h + o(h))(1 - \mu h - o(h))^{s-1} = s\mu h + o(h).$$

The probability that in this interval two or more servers complete a service is of the order $o(h)$. The same is true for the probability that in this interval one or more arrivals take place and at the same time one or more services are completed.

If the state i denotes the number of customers who are in the system at time t, either being served at a till or waiting in line, then obviously $\min(i, m)$ servers are busy. One arrival corresponds to a jump from i to $i + 1$ and one completion of service corresponds to a jump from i to $i - 1$. Since such jumps are the only ones possible we are here dealing with a *birth and death process*.

Let us consider the passages from i to j during an interval $[t, t + h)$. If $j \neq i + 1, j \neq i - 1$, then two or more jumps must have occurred. The probability for this is, according to the above, equal to $o(h)$. If $j = i + 1$, either one arrival has taken place but no further jumps have, or several jumps have occurred (e.g. two arrivals and one departure). Therefore we have

$$p_{ii+1}(h) = (\lambda h + o(h))(1 - \min(i, m)\mu h - o(h)) + o(h) = \lambda h + o(h).$$

If $j = i - 1$, then either exactly one service has been completed and no further jumps have taken place, or two or more jumps have taken place. Therefore we have in this case

$$p_{ii-1}(h) = (\min(i, m)\mu h + o(h))(1 - \lambda h - o(h)) + o(h)$$

$$= \min(i, m)\mu h + o(h).$$

If finally $j = i$ either no jump has taken place or at least two jumps have taken place and therefore we have

$$p_{ii}(h) = (1 - \lambda h - o(h))(1 - \min(i, m)\mu h - o(h)) + o(h)$$

$$= 1 - (\lambda + \min(i, m)\mu)h + o(h).$$

By comparing the general birth and death process with Example (b) we note that in this particular case $\lambda_i = \lambda$ and $\mu_i = \min(i, m)\mu$. Correspondingly the *backward equations* of Example (b) specialize and become

$$\dot{p}_{0j} = -\lambda p_{0j} + \lambda p_{1j},$$

$$\dot{p}_{ij} = -(\lambda + i\mu)p_{ij} + \lambda p_{i+1j} + i\mu p_{i-1j}, \quad 0 < i < m,$$

$$\dot{p}_{ij} = -(\lambda + m\mu)p_{ij} + \lambda p_{i+1j} + m\mu p_{i-1j}, \quad m \leqslant i,$$

and the *forward equations* become

$$\dot{p}_{i0} = -p_{i0}\lambda + p_{i1}\mu,$$

$$\dot{p}_{ij} = -p_{ij}(\lambda + j\mu) + p_{ij-1}\lambda + p_{ij+1}(j + 1)\mu, \quad 0 < j < m,$$

$$\dot{p}_{ij} = -p_{ij}(\lambda + m\mu) + p_{ij-1}\lambda + p_{ij+1}m\mu, \quad m \leqslant j. \qquad \square$$

(*d*) *Machine maintenance*. Suppose there are m machines which, if they are in use, can suffer a failure. For the repair of these machines in case of a failure $n \leqslant m$ mechanics are available. As soon as there is a failure of a machine it is repaired by one of the mechanics if he is available. Otherwise the machine has to wait.

We make similar assumptions to the ones in Example (*c*). If a machine has no breakdown at time t let the probability that the machine breaks down during the interval $[t, t + h)$ be equal to $\lambda h + o(h)$. If a mechanic is busy with the repair of a machine at time t let the probability that the repair job will be finished in the course of the interval $[t, t + h)$ be equal to $\mu h + o(h)$, while the probability that the mechanic will deal with two or more repairs during this period is equal to $o(h)$.

Let state i denote the number of machines out of order at time t. Then $\min(i, n)$ mechanics are busy and $m - i$ machines are working. Again only jumps from i to $i + 1$ (breakdown of a machine) and to $i - 1$ (completion of repair of a machine) are possible. Entirely analogously to Example (*c*) we have

$$p_{ii+1}(h) = (m - i)\lambda h + o(h),$$

$$p_{ii-1}(h) = \min(i, n)\mu h + o(h),$$

$$p_{ii}(h) = 1 - ((m - i)\lambda + \min(i, n)\mu)h + o(h).$$

We thus have a *birth and death process* with $\lambda_i = (m - i)\lambda$ and $\mu_i = \min(i, n)$.

We restrict ourselves to stating the *forward equations* for the initial state $i = 0$ where $p_{0j} = p_j$:

$$\dot{p}_0 = -p_0 m\lambda + p_1\mu,$$

$$\dot{p}_j = -p_j((m - j)\lambda + j\mu) + p_{j-1}(m - j + 1)\lambda + p_{j+1}(j + 1)\mu, \quad 0 < j < n,$$

$$\dot{p}_j = -p_j((m - j)\lambda + n\mu) + p_{j-1}(m - j + 1)\lambda + p_{j+1}n\mu, \quad n \leqslant j < m,$$

$$\dot{p}_m = -p_m n\mu + p_{m-1}\lambda, \quad p_j = 0 \text{ for } j > m. \qquad \square$$

If a *Markov chain* is given by a differential matrix determined by q_i, q_{ij}, we want to associate with it a *system with exponentially distributed durations* as follows: if the system has reached a state $i \in I$ at time s, then the duration of the state i up to the next jump is *exponentially* distributed with parameter q_i and *independent* of the system's past. With the next jump the system enters a state $j \neq i$ with probability $p_{ij} = q_{ij}/q_i$. Thus (p_{ij}) is a stochastic transition matrix with $p_{ii} = 0$. If we only consider the times of the jumps, then the states of the system immediately after the jumps can be considered as a Markov chain with *discrete parameter* and transition matrix (p_{ij}).

Suppose $s_{ij}(t)$ is the probability that the system with exponentially distributed durations is in the state j at time t, given that the system was in state i at time $t = 0$.

Obviously $s_{ii}(0) = 1$ and $s_{ij}(0) = 0$ for $i \neq j$, i.e. we have the same initial condition as for the transition probability functions $p_{ij}(t)$ of the Markov chain. If $j \neq i$ there must have been a *first* jump at a time $s \leqslant t$. The time s has the exponential density $q_i e^{-q_i s}$. The jump goes to a state k with probability q_{ik}/q_i and the probability that the system changes from k to j in the remaining time $t - s$ is $s_{kj}(t - s)$. We therefore have

$$s_{ij}(t) = q_i \int_0^t e^{-q_i s} \sum_{k \neq i} \frac{q_{ik}}{q_i} s_{kj}(t - s) \, ds. \tag{18}$$

If $j = i$ we have to consider in addition the possibility that the system has made no jump up to time t for which the probability is equal to $e^{-q_i t}$. Therefore we have

$$s_{ii}(t) = e^{-q_i t} + q_i \int_0^t e^{-q_i s} \sum_{k \neq i} \frac{q_{ik}}{q_i} s_{ki}(t - s) \, ds. \tag{19}$$

If we introduce in the integrals (18) and (19) the substitution $r = t - s$ and then differentiate with respect to t, we see that the $s_{ij}(t)$ are solutions of the backward equations (12). If these equations have *unique* solutions we must have $s_{ij}(t) = p_{ij}(t)$. We can thus *identify* the Markov chain with the associated system with exponentially distributed durations. The uniqueness is assured for chains with *finite* state space. The question of the uniqueness of the solutions of backward equations will be considered again in Section 6.3.

Examples

(*e*) *Poisson process.* In Example (*a*) the process remains in state i for an arbitrary period of time which is exponentially distributed with parameter λ. This corresponds to the exponentially distributed *renewal time* between two successive events (renewals) if the Poisson process is considered as a *renewal process* (as in Example (2.1.*a*)). □

(*f*) *Queueing systems.* The arrival process in Example (*c*) is a Poisson process. The time intervals between two successive arrivals are exponentially distributed with parameter λ. According to the elementary renewal theorem (the corollary to Theorem 2.3.2), the expected mean number of arrivals per unit time is equal to λ, since the exponential distribution with parameter λ has the expected value $1/\lambda$ (see Example $(1.3.g)$). The parameter λ is called the *rate of arrival*. If we assume that a server can serve continually then this serving process forms a Poisson process; the serving times in the system of Example (c) must therefore also be exponentially distributed with parameter μ. The parameter μ is thus called the *serving rate*.

Furthermore the queueing system of Example (c) always remains in a state i for an arbitrary period of time which is exponentially distributed with parameter $\lambda + i\mu$ or $\lambda + m\mu$ depending on whether $i < m$ or $i \geqslant m$. Of particular interest are the durations in the state 0, i.e. the *free* times of the system. They are exponentially distributed with parameter λ. $\qquad\square$

(g) *Machine maintenance.* Analogously to Example (f) it follows for the system of Example (d) that the failure-free time of the company's machines must be exponentially distributed with parameter λ, while the repair times are exponentially distributed with parameter μ. Furthermore the unoccupied times of the service crew, i.e. the length of stay in the state 0, are exponentially distributed with parameter $m\lambda$. $\qquad\square$

3.4. Limit distributions for chains with continuous parameter

The considerations concerning the classification of states of chains with discrete parameter can easily be applied to the continuous case. A state *i leads* to a state j, $i \Rightarrow j$, if there exists a $t > 0$ such that $p_{ij}(t) > 0$. Two states *communicate*, $i \Leftrightarrow j$, if $i \Rightarrow j$ and $j \Rightarrow i$. This leads, precisely as in the discrete case (see Section 3.2), to a *classification* of states which lead to themselves. Equally a set C is *closed* if there exists no state in C which leads to a state outside C; the chain is *irreducible* if the state space itself forms a class, which is necessarily *closed*. In order to simplify the presentation we shall restrict ourselves in the following to *irreducible chains*. It is left to the reader to extend the following considerations to the general case in an analogous way to that of Section 3.2.

As for the discrete case we call a distribution $\{p_i, i \in I\}$ *stationary* if we have

$$p_i = \sum_k p_k p_{ki}(t) \tag{1}$$

for all $t \geqslant 0$. If a stationary distribution p_i is an *initial distribution* of a Markov chain, then $p_i(t) = p_i$ for all $t \geqslant 0$. A comparison with $(3.3.17)$ shows that a stationary distribution has to satisfy the equations

$$p_i q_i = \sum_{j \neq i} p_j q_{ji}, \quad i \in I. \tag{2}$$

Conversely every distribution u_i, which is a solution of (2), is a stationary distribution. Indeed, if $p_i(t)$ is the absolute distribution at time t for the initial distribution μ_i, we obtain by differentiating (3.3.6) and by the *backward equations* (3.3.12)

$$\dot{p}_i = \sum_k u_k \dot{p}_{ki} = -\sum_k u_k q_k p_{ki} + \sum_k u_k \sum_{j \neq k} q_{kj} p_{ji}.$$

According to our assumption u_k satisfies (2) and it thus follows that $\dot{p}_i = 0$. Therefore $p_i(t) = $ constant $= u_i$ and u_i is indeed a stationary distribution.

The following theorem extends some results for discrete chains to the continuous case.

Theorem 1

For an *irreducible* continuous Markov chain exactly one of the following two statements holds:

(i) For all $i, j \in I$ we have

$$\lim_{t \to \infty} p_{ij}(t) = 0, \tag{3}$$

and there is *no* stationary distribution.

(ii) For all $i, j \in I$ we have

$$\lim_{t \to \infty} p_{ij}(t) = u_j > 0, \tag{4}$$

and $\{u_j, j \in I\}$ is the *only* stationary distribution.

Proof. It follows from (3.3.5) that

$$p_{ii}(s)p_{ik}(t) \leqslant p_{ik}(s+t) = p_{ii}(s)p_{ik}(t) + \sum_{j \neq i} p_{ij}(s)p_{jk}(t)$$

$$\leqslant p_{ik}(t) + (1 - p_{ii}(s)). \tag{5}$$

Since $p_{ii}(s) \to 1$ as $s \to 0$ we obtain $1 - \epsilon \leqslant p_{ii}(s) \leqslant 1$ for sufficiently small s. On account of (5) it thus follows that $|p_{ik}(t+s) - p_{ik}(t)| < \epsilon$ for sufficiently small s and the functions $p_{ik}(t)$ are *uniformly* continuous. Furthermore it follows for sufficiently small s that $p_{ik}(t+s) > 0$ if $p_{ik}(t) > 0$. Since $p_{ik}(t)$ cannot vanish for all t because the chain is irreducible, we get $p_{ik}(t) > 0$ for *all* sufficiently large t.

For $\delta > 0$ we now consider the *discrete* Markov chain with the *transition probabilities* $p_{ij}(\delta)$. For the n-step transition functions we obviously have $p_{ij}^{(n)}(\delta) = p_{ij}(n\delta)$. According to the above we have $p_{ij}(n\delta) > 0$ for all sufficiently large n, and the Markov chain is *irreducible* and *aperiodic*. By Theorems 3.2.4 and 3.2.9 we have as $n \to \infty$ either (i) $p_{ij}(n\delta) \to 0$ or (ii) $p_{ij}(n\delta) \to u_j > 0$. If δ' and δ'' are two *rational* numbers, then they will have infinitely many common multiples $n'\delta' = n''\delta''$. Therefore the limits of the $p_{ij}(n\delta)$ for rational δ *cannot* depend on δ. Owing to the uniform continuity of the $p_{ij}(t)$ we finally have $\lim p_{ij}(t) = \lim p_{ij}(n\delta)$ as $t \to \infty$ and $n \to \infty$ respectively.

If case (i) holds for discrete chains then these chains have to be transient or null-recurrent (see Theorems 3.2.4 and 3.2.5). By Theorem 3.2.9 there are no stationary distributions for discrete chains in case (i) and thus none for the continuous chain. In case (ii), however, there exists a unique stationary distribution u_i for the discrete chain corresponding to δ which satisfies (1) for all $t = n\delta$, by Theorem 3.2.9. Since δ is arbitrary, u_i satisfies (1) for *all* t and is thus a stationary distribution for the continuous chain. The theorem has thus been proved. ∎

Examples

(a) *Queueing systems.* We shall continue from Example (3.3.c). If the Markov chain introduced there has a stationary distribution $p_i, i = 0, 1, 2, \ldots$, then this distribution must, by (2), be the solution of the equations

$$\lambda p_0 = \mu p_1,$$

$$(\lambda + j\mu)p_j = \lambda p_{j-1} + (j + 1)\mu p_{j+1}, \quad 0 < j < m,$$

$$(\lambda + m\mu)p_j = \lambda p_{j-1} + m\mu p_{j+1}, \quad m \leq j.$$

If we subtract every $(j - 1)$th from the jth equation, we obtain a new system which can easily be solved recursively. We find

$$p_j = p_0 \frac{(\lambda/\mu)^j}{j!}, \quad 0 \leq j < m,$$

$$p_j = p_0 \frac{(\lambda/\mu)^j}{m! m^{j-m}}, \quad m \leq j.$$

We have

$$\sum_{j=0}^{\infty} p_j = p_0 \left\{ \sum_{j=0}^{m-1} \frac{(\lambda/\mu)^j}{j!} + \sum_{j=m}^{\infty} \frac{(\lambda/\mu)^j}{m! m^{j-m}} \right\}.$$

The series converges if $\lambda/\mu < m$ and diverges for $\lambda/\mu \geq m$. The chain is in both cases *irreducible.* There exists a stationary limit distribution if $\lambda/\mu < m$ but not otherwise. If $\lambda/\mu \geq m$ then $p_j(t) \to 0$ as $t \to \infty$ for all j (Theorem 1). The parameter λ is the mean *rate of arrival* (mean number of arrivals per unit time) and μ the mean *serving rate* of a till (mean number of services per unit time (see Example (3.3.f))). The serving rate of the m tills together is $m\mu$. If the rate of arrival λ is greater than the serving rate $m\mu$ of the system, the number of customers in the system will grow unboundedly with probability 1 during the course of time. This is called an *unstable* system. If $\lambda < m\mu$, we can normalize the series of p_j to 1 by setting

$$p_0 = \frac{1}{\sum_{j=0}^{m-1} \dfrac{(\lambda/\mu)^j}{j!} + \sum_{j=m}^{\infty} \dfrac{(\lambda/\mu)^j}{m!\, m^{j-m}}} = \frac{1}{\sum_{j=0}^{m-1} \dfrac{(\lambda/\mu)^j}{j!} + \dfrac{(\lambda/\mu)^m}{m!\,(1-(\lambda/m\mu))}}.$$

If we replace p_j in the above formulae by p_0 we obtain a stationary distribution. According to Theorem 1 and (3.3.6) the absolute distributions $p_j(t)$ converge for any arbitrary initial distribution to the stationary distribution p_j as $t \to \infty$.

The cases $m = 1$ (one server only) and $m = \infty$ (unlimited serving capacity) are particularly simple. For $m = 1$ it follows from the above results that

$$p_0 = \frac{1}{1 + \dfrac{(\lambda/\mu)}{1-(\lambda/\mu)}} = 1 - \frac{\lambda}{\mu},$$

$$p_j = \left(\frac{\lambda}{\mu}\right)^j \left(1 - \frac{\lambda}{\mu}\right), \quad \frac{\lambda}{\mu} < 1.$$

This is a geometric distribution with $p = \lambda/\mu$ (see Example $(1.2.c)$). The expected number of customers in the system is for the stationary limit distribution equal to (see Example $(1.3.d)$)

$$\sum_{j=0}^{\infty} j p_j = \frac{(\lambda/\mu)}{1-(\lambda/\mu)}.$$

For $m = \infty$ on the other hand we obtain

$$p_0 = \frac{1}{\sum_{j=0}^{\infty} \dfrac{(\lambda/\mu)^j}{j!}} = e^{-\lambda/\mu},$$

$$p_j = e^{-\lambda/\mu} \frac{(\lambda/\mu)^j}{j!}, \quad \frac{\lambda}{\mu} \text{ arbitrary}.$$

This is a Poisson distribution. The mean number of customers in the system is in this case equal to the expected value of the Poisson distribution which is λ/μ (see Example $(1.3.e)$). This value is, for $\lambda/\mu < 1$, smaller than the corresponding expected value in the case $m = 1$. For $1 < m < \infty$ the expected values of the number of customers in the system lie between these bounds. $\square$

(b) *Machine maintenance.* This example continues from Example $(3.3.d)$. A stationary distribution has to satisfy the equations

$$m\lambda p_0 = \mu p_1,$$

$$((m-j)\lambda + j\mu)p_j = (m-j+1)\lambda p_{j-1} + (j+1)\mu p_{j+1}, \quad 0 < j < n,$$

$$((m-j)\lambda + n\mu)p_j = (m-j+1)\lambda p_{j-1} + n\mu p_{j+1}, \quad n \leqslant j < m,$$

$$\lambda p_{m-1} = n\mu p_m.$$

We obtain recursively from the first equation $p_1/p_0 = m\lambda/\mu$, from the second

$$(j + 1)\mu p_{j+1} = (m - j)\lambda p_j$$

and from the third

$$n\mu p_{j+1} = (m - j)\lambda p_j.$$

The fourth equation is redundant. From this the ratios p_j/p_0 can be computed and p_0 can be determined from the normalization condition $p_0 + p_1 + \ldots + p_m = 1$. We have therefore always a stationary distribution. The case of one mechanic ($n = 1$) is again particularly simple. The reader is advised to examine this case further and to determine in particular the expected number of machines which have broken down.   □

(c) *Birth and death processes.* These processes were introduced in Example (3.3.b). If there exists a stationary distribution $p_i, i = 0, 1, 2, \ldots$, it has to satisfy the equations

$$\lambda_0 p_0 = \mu_1 p_1,$$

$$(\lambda_j + \mu_j)p_j = \lambda_{j-1}p_{j-1} + \mu_{j+1}p_{j+1}, \quad j > 0.$$

From this system it follows recursively that

$$\lambda_j p_j = \mu_{j+1} p_{j+1}, \quad j \geqslant 0,$$

and therefore

$$p_j = \frac{\lambda_0 \lambda_1 \ldots \lambda_{j-1}}{\mu_1 \mu_2 \ldots \mu_j} p_0.$$

These exists therefore a stationary distribution if and only if the series

$$\sum_{j=1}^{\infty} \frac{\lambda_0 \lambda_1 \ldots \lambda_{j-1}}{\mu_1 \mu_2 \ldots \mu_j}$$

converges. For pure birth processes the equations have only the trivial solution $p_i = 0$. There exists no stationary distribution.   □

3.5. Computation of transient solutions of continuous parameter Markov chains

The limiting behaviour of an irreducible continuous parameter Markov chain is uniquely determined by its stationary distribution if it exists (see Theorem 3.4.1). Its numerical computation amounts to the solution of a system of linear equations (3.4.2). But there are many cases where one is much more interested in the behaviour of the chain for finite times t rather than for $t \to \infty$. This is the case for irreducible *transient* Markov chains and also for Markov models where the class of *transient* states is of most interest (see Example (*a*) below). Even in the case of an irreducible *recurrent* chain one may be interested in the time-dependent manner in which the limiting stationary distribution is approached.

This demands the numerical computation of the *transition probabilities* $p_{ij}(t)$ from the Kolmogorov equations (3.3.12) or (3.3.13). These are systems of ordinary linear differential equations which have to be solved for initial conditions $p_{ij}(0) = \delta_{ij}$ (3.3.11). At least for finite state spaces one could therefore apply Laplace-transform techniques together with eigenvalue analysis, as described in most textbooks on differential equations. This would be an approach analogous to Example (3.1.*g*). But this is of more theoretical than practical value as the computations of eigenvalues is rather awkward except in some simple cases. Hence a solution of the Kolmogorov equations by numerical methods such as the methods of Euler or Runge–Kutta is more practical. But these general numerical methods do not take into account the *particular* structure of the Kolmogorov equations which is induced by the fact that the q_{ij}, $-q_i$ form a *differential* matrix A. Therefore a tailored method will be described below which is motivated by the probabilistic content of the problem.

A more specific problem than the computation of the transition probabilities $p_{ij}(t)$ is the computation of the *absolute distribution* $p_i(t)$ for some initial distribution $p_i(0)$. This is done by solving equations (3.3.17). Of course this problem is closely related to the original problem (3.3.13).

Examples

(*a*) *Reliability of k-out-of-n systems with repair.* Suppose a system is composed of n identical modules and the system is to be considered as functioning only if *at least* k out of the n modules are functioning. Suppose furthermore that there are s repair channels (repairmen) of which each can repair one module at a time. This is a repair organization just as in Example (3.3.*d*). And just as there, let state i be the number of modules not functioning, $\lambda h + o(h)$ the probability that a functioning module breaks down in an interval $[t, t + h)$ and $\mu h + o(h)$ the probability that a busy repair channel completes a repair during the interval $[t, t + h)$.

In contrast to Example (3.3.*d*) only states $i = 0, 1, \ldots, n - k, n - k + 1$ are considered. State $n - k + 1$ is the *down state* of the system, because in this state the minimal number k of functioning modules is no longer available. This state will be made *absorbing*. Otherwise one has the same *birth and death process* as in Example (3.3.*d*) with $\lambda_i = (n - i)\lambda$ and $\mu_i = \min(i, s)$ for $i = 0, 1, \ldots, n - k$, $\mu_{n-k+1} = 0$ (note that state $n - k + 1$ is in fact absorbing). This is the case referred to as *hot reserve*, because all functioning modules are supposed to be 'on' and hence candidates for breakdown. An alternative is *cold reserve*, where at any time only k functioning modules are 'on' and only these ones are candidates for breakdown. In this case $\lambda_i = k\lambda$ for $i = 0, 1, \ldots, n-k$.

In either case the states $i = 0, 1, \ldots, n-k$ form a *transient* class. The stationary distribution of the chain exists but is of no interest; it is simply $p_i = 0$ for $i = 0, 1, \ldots, n-k$, $p_{n-k+1} = 1$. On the other hand let $p_0(0) = 1$, $p_i(0) = 0$ for $i = 1, 2, \ldots, n-k+1$ such that the system starts entirely intact and let $p_i(t)$, $t > 0$, be the absolute distribution associated with this initial distribution. Then $p_{n-k+1}(t)$ is the probability of a system breakdown *before* time t, and hence $p_{n-k+1}(t)$ is the *distribution function of the time until the first system failure.*

Alternatively let $p_{n-k}(0) = 1$ and $p_i(0) = 0$ otherwise. This is the initial condition of a system which is reset to operation after failure as soon as k functioning modules become available. In this case $p_{n-k+1}(t)$ is the *distribution function of the time between two failures.*

The computation of both distributions depends on the computation of the transient solutions of the underlying Markov chain.   □

(b) *Waiting time in a queueing system.* Consider the queueing system of Example (3.3.c), but with only one server, i.e. with $m = 1$. What is the distribution of the waiting time of a customer C arriving at some time s in this system? The answer depends on the queueing discipline, i.e. how the next customer in the queue is chosen for service each time the server becomes free. Let us assume a *random order*: the customer to be served is chosen at random with equal probabilities among the customers in the queue.

The problem can be formulated in terms of a continuous Markov chain. States $i = 1, 2, \ldots$ signify i customers waiting for service in the queue. State 0 means service of customer C has already started; it will be an *absorbing* state. The time s of arrival of C in the system will be the time origin of the process. So if at time $s + t$ the process is in state 0 this means that the waiting time of C was smaller than t. Hence $p_0(t)$ will be the *distribution function of the waiting time of C.*

The following jumps are possible for $i = 1, 2, \ldots$

$i \rightarrow i + 1$ arrival of a new customer,

$i \rightarrow i - 1$ beginning of service of a customer different from C,

$i \rightarrow 0$ beginning of service of customer C.

Consider an interval $[t, t + h)$. The probability of a new arrival is, as in Example (3.3.c), equal to $\lambda h + o(h)$, the probability of the end of a service and hence the beginning of a new one equals $\mu h + o(h)$. In state i there is a chance of $1/i$ that this new service will concern customer C. Thus it follows that $q_{ii+1} = \lambda$, $q_{i0} = \mu/i$, $q_{ii-1} = (i-1)\mu/i$ and $q_i = (\lambda + \mu)$. All other $q_{ij} = 0$, especially $q_{0j} = 0$ since 0 is absorbing.

The *initial distribution* of this process is the *absolute distribution* of the process of Example (3.3.c) at time s. If this process has a stationary distribution, then for large s this stationary distribution can be taken as the initial distribution (see Example (3.4.a)). The waiting time distribution $p_0(t)$ is obtained by computing the transient solution of the above Markov chain. □

The Kolmogorov equations (3.3.12) and (3.3.13) in matrix notation are

$$\dot{P} = AP \quad \text{and} \quad \dot{P} = PA, \tag{1}$$

where $\dot{P}$ denotes the matrix with elements $\dot{p}_{ij}$. A is the differential matrix of the Markov chain and P the matrix of the transition probabilities (see Section 3.3). The following theorem of matrix theory is used to solve (1).

Theorem 1
If $A = (a_{ij})$, $i, j \in I$, is a real matrix with

$$\|A\| = \sup_{i \in I} \left(\sum_{j \in I} |a_{ij}| \right) < \infty \tag{2}$$

then the series

$$\sum_{n=0}^{\infty} \frac{A^n t^n}{n!}, \tag{3}$$

with $A^0 = I$ (identity matrix), converges (element-wise) absolutely for all real t.

Proof. Let C and B be two matrices with $\|C\|, \|B\| < \infty$. Then

$$\left| \sum_k \left| \sum_j b_{ij} c_{jk} \right| \right| \leqslant \sum_k \sum_j |b_{ij}| \, |c_{jk}|$$

$$= \sum_j |b_{ij}| \sum_k |c_{jk}| \leqslant \sum_j |b_{ij}| \, \|C\|$$

$$\leqslant \|B\| \, \|C\|.$$

This implies that $\|BC\| \leqslant \|B\| \, \|C\|$ and especially $\|A^n\| \leqslant \|A\|^n$, $n = 1, 2, \dots$. Now each element of (3) has the form

$$\sum_{n=0}^{\infty} a_{ij}^{(n)} \frac{t^n}{n!}, \tag{4}$$

where $a_{ij}^{(n)}$ is an element of A^n. But $|a_{ij}^{(n)}| \leqslant \|A^n\| \leqslant \|A\|^n$. Hence the series (4) is majorized by the absolutely convergent series

$$\sum_{n=0}^{\infty} \|A\|^n \frac{t^n}{n!} = e^{\|A\|t}. \tag{5}$$

This proves the theorem. ■

It is natural to put

$$\sum_{n=0}^{\infty} \frac{A^n t^n}{n!} = e^{At}, \tag{6}$$

and it is easily verified that (assuming $\|A\|, \|B\| < \infty$)

$$e^{(A+B)t} = e^{At} e^{Bt}, \tag{7}$$

$$e^{A(t+s)} = e^{At} e^{As}, \tag{8}$$

$$\frac{\mathrm{d}}{\mathrm{d}t} e^{At} = A e^{At} = e^{At} A. \tag{9}$$

This matrix exponential function represents in most cases the solution of the Kolmogorov equations.

Theorem 2

Let A be the differential matrix of a Markov chain. If the q_i are bounded, then

$$P(t) = e^{At} \tag{10}$$

is a solution to the Kolmogorov equations (1), with $P(0) = I$.

Proof. The theorem follows immediately from (9).   ∎

Note that this is an existence theorem. It will be shown later in Section 6.3 that under the condition of the theorem (boundedness of the q_i) this solution is *unique*. But we have yet to verify that the $P(t)$ given by (10) are in fact transition probabilities. By (8) $P(t)$ satisfies the Chapman–Kolmogorov equations (3.3.5). But is $P(t)$ a *stochastic* matrix for all $t \geqslant 0$? The next theorem asserts this. Its proof also contains the key for a convenient numerical computation procedure for $P(t)$.

Theorem 3

Under the conditions of Theorem 2, $P(t)$ is a *stochastic* matrix.

Proof. Let $c \geqslant \sup_{i \in I} q_i$ and define the matrix

$$Q = I + (1/c)A. \tag{11}$$

Q is a *stochastic* matrix, because A is a differential matrix. Now

$$\begin{aligned}
P(t) = e^{At} &= e^{(I + (1/c)A)ct - Ict} \\
&= e^{Qct} e^{-ct} \\
&= \sum_{n=0}^{\infty} Q^n \frac{(ct)^n}{n!} e^{-ct}.
\end{aligned} \tag{12}$$

Q is a transition matrix of a discrete Markov chain, the Q^n are its n-stage transition matrices. Suppose that the times between two consecutive transitions of this chain are exponentially distributed with parameter c. Then the number of transitions from time 0 to t just follows the Poisson distribution appearing in (12). This gives (12) a probabilistic interpretation as transition probabilities of this special process. Hence the elements of $P(t)$ must in fact be probabilities and $P(t)$ must be stochastic. ∎

The representation (12) of $P(t)$ is much more suited for numerical computation than (10). The elements of the series (10) are given in (4). For practical calculations the infinite series have to be truncated at a sufficiently large m. Then (4) becomes a *polynomial* in t, which for fixed m grows in absolute value without bounds as $t \to \infty$. This indicates the kind of difficulty which one encounters in using (4). For large t one has in (4) to calculate differences of large numbers, with the well-known numerical problems associated with this (cancellation of decimal places in numbers with a finite number of decimal places) which very quickly render calculations senseless.

On the other hand all terms of the series (12) are *positive* and no such numerical difficulties arise. The truncation error can easily be estimated. Let

$$P(t) = \sum_{n=0}^{m} Q^n \frac{(ct)^n}{n!} e^{-ct} + R^{(m)}(t). \tag{13}$$

The elements of the error matrix $R^{(m)}(t)$ are

$$0 \leqslant r_{ij}^{(m)}(t) = \sum_{n=m+1}^{\infty} q_{ij}^{(n)} \frac{(ct)^n}{n!} e^{-ct}$$

$$\leqslant \sum_{n=m+1}^{\infty} \frac{(ct)^n}{n!} e^{-ct}. \tag{14}$$

Thus the truncation errors in the series (12) are bounded by the tail of the Poisson distribution. This also shows that it is convenient to choose c as small as possible, but greater than all q_i (see the beginning of the proof of Theorem 3); for example, $c = \sup q_i$.

To compute the absolute distribution $p(t)$ associated with an initial distribution $p(0)$ note that $p(t) = p(0)P(t)$. Hence it follows from (12) that

$$p(t) = \sum_{n=0}^{\infty} p(0)Q^n \frac{(ct)^n}{n!} e^{-ct}$$

$$= \sum_{n=0}^{\infty} q^{(n)} \frac{(ct)^n}{n!} e^{-ct}, \tag{15}$$

where the $q^{(n)} = p(0)Q^n$ are the absolute distributions of the *discrete* parameter Markov chain with transition matrix Q. The $q^{(n)}$ can be computed recursively by

$$q^{(n)} = q^{(n-1)}Q, \quad n = 1, 2, \ldots . \tag{16}$$

This shows another advantage of the approach. In applications A often contains many zero entries; it is a *sparse* matrix. This carries over to Q. Special storage techniques can be used to store sparse matrices in order to save memory space. Using (16) no other matrices have to be stored.

Notes to Chapter 3

There exist numerous textbooks and monographs on Markov chains. [1] treats the subject of this chapter in an extensive mathematical representation. [7] and [8] also develop the theory of discrete parameter Markov chains very extensively, but from a slightly different angle. [3] is a recommendable, elementary, but very inspiring, introduction to some of the modern aspects of the theory of Markov processes.

Section 3.5 is based on [4], [5]. Other computational aspects of Markov chains are discussed in [2] and [6]. [2] also exhibits an important domain of application of Markov chains.

Markov chains can be generalized to *Markov processes* by introducing state spaces which are non-countable. In the case of continuous time parameter there are essentially two different classes of processes. If the system remains in a given state for an exponentially distributed time and then jumps to another state, one has a *jump process*. This is a direct generalization of the system with exponentially distributed sojourn times introduced in Section 3.3. If on the other hand the system changes its state continuously the process is called a *diffusion process*. For these more general processes see [1.3, vol. 2], [1.2] and [1.4].

Another important generalization of Markov chains are the *Markov renewal processes* or the *semi-Markov processes*, which will be introduced and discussed in Chapter 7. It is fair to say that most of the stochastic problems of operations research are treated by reducing them to Markov chains. This will be documented in the following chapters.

4

Queues

4.1. Queueing processes

Queues are a well-known sight in daily life. They appear, however, also in industrial production systems, in traffic and transport and especially in information transmission (telephone) and information processing systems (time sharing and real time computer systems). The purpose of queueing theory is to examine queueing phenomena systematically. It is considered to be one of the classic disciplines of operations research.

Queueing systems appear in a great number of variations. The system described in the following can be considered as a basic model. A number s of *servers* is available. Each server can only serve one *unit* at a time. These units are often called *customers*. The customers arriving in the system are numbered $i = 0, 1, 2, \ldots$. The serving period of a unit i by the server is generally random and is denoted by a random variable $B_i, i = 0, 1, 2, \ldots$. In this basic example the B_i are assumed to be mutually *independent* and we suppose that every serving period B_i has the same distribution function $F_B(t)$. The service times can only take *positive* values so that $F_B(t) = 0$ for $t \leqslant 0$. If a server is continually busy, then the successive services form a renewal process. It follows from the elementary renewal theorem, the corollary to Theorem 2.3.2, that the mean number of served units per unit time for one server is equal to $\mu = 1/E[B_i]$. The parameter μ is the *serving rate* of each individual server.

The units arrive individually and generally at irregular intervals in the system. Let the units be numbered in the order of their arrival. The time intervals between the $(i-1)$th and ith units are denoted by the random variables $A_i = 1, 2, \ldots$. Let these random variables be *mutually independent* as well as independent of the service time B_i and *identically distributed* with the distribution function $F_A(t)$. Of course, we again have $F_A(t) = 0$ for $t \leqslant 0$. The times of the arrivals thus form a *renewal* process. The parameter $\lambda = 1/E[A_i]$ is the *arrival rate*, i.e. the mean number of arrivals per unit time (see the elementary renewal theorem, the corollary to Theorem 2.3.2).

If a newly arrived unit finds a free server, then the service starts at once. If on the other hand all servers are busy at the arrival of a unit, the unit has to wait in the *queue*. As soon as a server is free, the next unit to be served is selected from the queue. This method is called *queueing discipline*. Generally best known is the so-called FIFO-discipline (First In, First Out), where the units are being served on a first come, first served basis. Other possibilities are the LIFO-discipline (Last In, First Out), where the last arrival is served first, or the random choice of a unit waiting for service from the units in the queue.

This basic example can be varied in many ways. For example, the units could arrive in groups instead of individually and be served as such. Furthermore the units could have varying priorities which determine the order in which they are served. These are only a few of the various possibilities.

Such multi-channel systems are denoted in the so-called *Kendall* notation by the notation $F_A/F_B/s$ which specifies (in that order) inter-arrival time distribution, service time distribution and the number of servers. In particular the symbols M for the *exponential distribution* and G (general) for arbitrary distributions that are not further determined have established themselves. If we want to stress that the inter-arrival times A_i are mutually independent as assumed above, we write GI (General Independent). $GI/G/s$ therefore denotes the general system introduced above with not especially determined inter-arrival and service times. $M/G/1$ denotes a system with *one* server with an arbitrary service distribution and *exponentially* distributed inter-arrival times. The arrival times form a *Poisson process* in this case; see Example $(2.1.a)$. The system introduced in Example $(3.3.c)$ emerges now as an $M/M/s$-system with the special cases $M/M/1$ and $M/M/\infty$ (see also Examples $(3.3.f)$, $(3.4.a)$).

A great many *stochastic processes* can be introduced in connection with the described queueing system. First let $X_t, t \geqslant 0$, be the number of units which are in the system at time t. Then X_t forms a *continuous Markov chain* in the case of the $M/M/s$-system, as can be seen from Example $(3.3.c)$. This is, unfortunately, not the case for more general systems (see Section 4.2). The *sample functions* of X_t can be interpreted as *step functions* with a jump of $+1$ at the arrival of one unit and a jump of -1 after a service has been completed. Note that the process X_t is *not* influenced by the queueing discipline (FIFO, LIFO or random choice).

Secondly let $W_i, i = 1, 2, \ldots$, be the *length of time in the queue* or *waiting time* of the ith customer in the system from his arrival up to the completion of the service. The distribution of the W_i depends, of course, on the queueing discipline, in contrast to the one of the X_t.

Thirdly there exist a number of *embedded renewal processes*. Particularly interesting times are those arrivals of units in a queueing system which find *all* servers unoccupied. The time interval between the arrival of two units which find all servers unoccupied is called a *cycle*. At this point it is assumed that at $t = 0$ a cycle starts, i.e. that at the beginning all servers are unoccupied and one unit arrives. We now let $Z_i, i = 1, 2, \ldots$, be the *cycle lengths*, i.e. the periods of time which elapse between cycles. Furthermore let $Y_i, i = 1, 2, \ldots$, be the *number of units served during the ith cycle*. Since at the beginning of each cycle the same stochastic mechanism begins to work, the Z_i as well as the Y_i are mutually *independent* and *identically distributed* (Z_i and Y_i by contrast are not independent). $\{Z_i, i = 1, 2, \ldots\}$ and $\{Y_i, i = 1, 2, \ldots\}$ thus each define a *renewal process*. Finally, let $V_i, i = 1, 2, \ldots$, be the *total waiting time* of all the Y_i units served during the cycle. $\{V_i, i = 1, 2, \ldots\}$ also forms a *renewal process*.

If the renewal process Z_i is *transient* there is a last unit which will find all servers unoccupied at its arrival (see Section 2.3). Any units that arrive after that will never find all servers free. It follows from this that the Y_i and V_i renewal processes are also *transient*. If the Z_i renewal process is *recurrent* the same will be true for the Y_i and V_i renewal processes. In this case let us call the cycles recurrent. The renewal theory now yields the proof of the following theorem:

Theorem 1

If the cycles are recurrent and if the expected values $E[Z_1], E[Y_1], E[V_1]$ are finite, we have

$$\lim_{t \to \infty} \frac{1}{t} E\left[\int_0^t X_t \, dt\right] = E[V_1]/E[Z_1], \tag{1}$$

$$\lim_{n \to \infty} \frac{1}{n} E\left[\sum_{i=1}^n W_i\right] = E[V_1]/E[Y_1], \tag{2}$$

$$E[Y_1]/E[Z_1] = \lambda. \tag{3}$$

Proof. The sequence $\{(V_i, Z_i), i = 1, 2, \ldots\}$ defines a *cumulative process*

$$\sum_{i=1}^{N(t)+1} V_i, \tag{4}$$

where $N(t)$ is the counting process of the renewal process of the Z_i; (see (2.3.23)). By Theorem 2.3.5 we have

$$\lim_{t \to \infty} E\left[\sum_{i=1}^{N(t)+1} V_i\right] \bigg/ t = E[V_1]/E[Z_1]. \tag{5}$$

(4) gives the sum of the waiting times of the units during the first $N(t) + 1$ cycles. If during the time interval from t up to $t + dt$ altogether X_t units are in the system, the sum of the waiting times of these units during this interval is equal to $X_t dt$. Summation (4) can thus also be obtained by integration of the step function X_t over the duration of the $N(t) + 1$ cycles, i.e. (4) is equal to

$$\int_0^{Z_1 + \ldots + Z_{N(t)+1}} X_t \, dt = \int_0^t X_t \, dt + \int_t^{Z_1 + \ldots + Z_{N(t)+1}} X_t \, dt. \tag{6}$$

If we take the expected value of (6), the expected value of the second term on the right hand side of (6) is at the most equal to $E[V_1]$ and thus vanishes after dividing by t for $t \to \infty$. (1) therefore follows from (5).

The sequence $\{(V_i, Y_i), i = 1, 2, \dots\}$ also defines a cumulative process. Since the Y_i are *arithmetic* we only look at this process for $n = 1, 2, \dots$. The cumulative process is defined by

$$\sum_{i=1}^{N(n)+1} V_i, \tag{7}$$

where $N(n)$ is the counting process for the discrete renewal process of the Y_i. It again follows from Theorem 2.3.5 that

$$\lim_{n \to \infty} E\left[\sum_{i=1}^{N(n)+1} V_i \middle| n \right] = E[V_1]/E[Y_1]. \tag{8}$$

From the definition of V_i (7) is equal to

$$\sum_{k=1}^{n} W_k + \sum_{k=n+1}^{Y_1+\dots+Y_{N(n)+1}} W_k. \tag{9}$$

If we take the expected value of this, the second term is at most equal to $E[V_1]$. Equation (2) therefore follows from (8).

Finally, the sequence $\{(Z_i, Y_i), i = 1, 2, \dots\}$ again defines a cumulative process

$$\sum_{i=1}^{N(n)+1} Z_i = \sum_{k=1}^{n} A_k + \sum_{k=n+1}^{Y_1+\dots+Y_{N(n)+1}} A_k, \tag{10}$$

where the A_k are the intervals between two arrivals. It follows from Theorem 2.3.5 that

$$\frac{E[Z_1]}{E[Y_1]} = \lim_{n \to \infty} \frac{1}{n} E\left[\sum_{k=1}^{n} A_k \right] = E[A_1] = \frac{1}{\lambda} \tag{11}$$

which corresponds to (3).   ∎

The integral

$$\frac{1}{t} \int_0^t X_t \mathrm{d}t$$

is the number of units in the system during the interval $[0, t]$ averaged over time. Correspondingly the left hand side of (1) represents the expected value of the average number of units in the system averaged over the infinite length of time. This value is denoted by L (average queue length). The left hand side of (2) represents the expected value of the waiting times averaged over the units served. Let us denote this value by W. It follows from Theorem 1 that

$$L = \lambda W \tag{12}$$

or, in words: *average queue length = rate of arrival × average waiting time.*
Equation (12) is a form of the $L = \lambda W$ law which occurs in many variations
in queueing theory. If, for example, the limits of $E[W_i]$ and $E[X_t]$ exist for
$i \to \infty$ and $t \to \infty$ respectively, then it is obvious that $\lim E[W_i] = W$ and
plausible (but still to be proved) that $\lim E[X_t] = L$. This result is another
form of the law $L = \lambda W$. We can conclude from (12) in particular that the
average waiting time W is *independent of the queueing discipline* since the
process X_t and therefore L are independent of it!

Example

   (a) $L = \lambda W$ *is the law for the M/M/∞- and M/M/1-systems.* For the
$M/M/\infty$-system the waiting time of every unit is equal to its service time
which is exponentially distributed with parameter μ; see Example (3.3.*f*).
Let W be the expected value of the waiting time, $W = 1/\mu$. Let L be the
expected value of the number of units in the system under the stationary
limit distribution. According to Example (3.4.*a*), $L = \lambda/\mu$. With this definition of L and W we obtain the relation $L = \lambda W$.

   For the $M/M/1$-system we may consider a test unit which encounters on
arrival X units in the system, where X is distributed according to the *stationary limit distribution* $P[X = j] = p_j, j = 0, 1, 2, \ldots$ of Example (3.4.*a*). Let
T be the waiting time of the test unit from its *arrival* up to the *moment when
it gets served.* If $X = 0$, the waiting time $T = 0$ and thus, if $\rho = \lambda/\mu$,

$$P[T = 0] = p_0 = 1 - \rho.$$

If $X = j$, the waiting time of the test unit is equal to the sum of the service
times of the j previously arrived units (with the FIFO-discipline). Thus, by
Example (2.1.*a*) we have

$$P[T > t \mid X = j] = e^{-\mu t} \sum_{k=0}^{j-1} \frac{(\mu t)^k}{k!}, \quad j = 1, 2, \ldots .$$

It follows that

$$P[T > t \mid X > 0] = \sum_{j=1}^{\infty} P[T > t \mid X = j] P[X = j \mid X > 0] = e^{-(1-\rho)\mu t}.$$

The conditional waiting time, given that the system is not empty on arrival,
has an exponential distribution and the corresponding expected value of the
waiting time $E[T \mid X > 0]$ is equal to $1/(1 - \rho)\mu$.

   Let $W' = E[T]$ be the expected value of the waiting time up to the
beginning of service

$$W' = E[T] = E[T \mid X > 0] P[X > 0] = \rho/(1 - \rho)\mu.$$

Let W'' be the expected value of the queueing time from the time of arrival up to completion of the service, $W'' = W' + 1/\mu = 1/(1 - \rho)\mu$. Furthermore, let L'' be the expected value of the number of units in the system under the stationary limit distribution, $L'' = \rho/(1 - \rho)$; see Example (3.4.*a*). Finally let L' be the expected value of the number of units in the queue without the unit which is being served, $L' = L'' - \rho = \rho^2/(1 - \rho)$. Thus we have the $L = \lambda W$ laws: $L' = \lambda W'$ and $L'' = \lambda W''$. $\qquad\qquad\square$

4.2. Embedded Markov chains

For practical applications we may often assume for good reasons that the arrivals form a Poisson process. Less well justified on the other hand in many cases is the assumption that we have an exponentially distributed service time. What is of interest, however, is to examine the $M/G/1$-system in order to test the sensitivity of the results for the $M/M/1$-system under deviation from the exponential distribution of the service time. This analysis of the $M/G/1$-system will enable us at the same time to introduce the method of the *embedded Markov chain* which can also be applied to other systems.

As in Section 4.1 let X_t be the number of units which are in the system at time t. The knowledge of X_t is generally not sufficient to determine the development of the process X_{t+s} for $s \geq 0$, i.e. past the time t. If $X_t \geq 1$, there is a unit being served at time t. This service will be completed after a time s' and at time $t + s'$ $X_{t+s'}$ will make one jump down since a unit whose service has been completed leaves the system. The distribution of the time s', however, depends on the *moment at which the service has begun*. If r' stands for the length of service with distribution function $P[r' \leq x] = G(x)$ and if the service at time t has already lasted for t' unit times, we have $s' = r' - t'$. The distribution of s' is equal to the *conditional* distribution of r', given that $r' \geq t'$,

$$P[s' > x | r' > t'] = P[r' > x + t' | r' > t'] = \frac{1 - G(x + t')}{1 - G(t')}. \qquad (1)$$

To determine X_{t+s}, $s \geq 0$, we must therefore not only know X_t but also at least the time at which the current service began. Therefore X_t is generally *not* a Markov chain.

If s'' is the time from t up to the next arrival, r'' is the interval between two arrivals and if t'' unit times have elapsed since the last arrival, it can be argued similarly that the distribution of s'' is equal to the conditional distribution of r'', given that $r'' \geq t''$. For the $M/G/1$-system r'' is *exponentially* distributed with parameter λ. In this case it follows from (1) that

$$P[s'' \geq x | r'' \geq t''] = e^{-\lambda(x+t'')}/e^{-\lambda t''} = e^{-\lambda x} \qquad (2)$$

and the distribution of s'' is, *independently* of t'', again precisely the same exponential distribution as the distribution of r''. No matter how long it has been since the last arrival, the time until the next arrival always has the same exponential distribution. In order to determine the next arrival it is therefore *not* necessary to know the time of the last arrival. For this also compare the distribution of the residual life for the Poisson process (Example (2.1.*b*)).

If the service time is also *exponentially* distributed, we have the same result. This is the reason why X_t is a Markov chain for $M/M/1$ (and more generally for the $M/M/s$-system). We can remedy the situation for the $M/G/1$-system, as well as for several other systems, by considering the X_t process not at all times but only at certain, *appropriately chosen*, discrete times. This is how we can regain the Markov property in many cases.

For the $M/G/1$-system it is advantageous to consider only *times immediately after the exit of a unit*. Let X_k be the number of units which remain in the system *immediately after* the exit of the kth unit. Furthermore let V_k be the number of units which arrive during the *next* service. If the kth unit leaves the system *empty* after its exit ($X_k = 0$), then before the exit of the next unit at least one unit has to have arrived, (the $(k + 1)$th), whose service starts immediately on arrival. At its exit this unit leaves behind the V_k units which have just arrived during its service. In this case we have $X_{k+1} = V_k$. If on the other hand at its exit the kth unit leaves behind at least one unit ($X_k > 0$), then the service of the $(k + 1)$th unit begins immediately. Up to the exit of this unit V_k new units enter the system, so that after subtracting the leaving unit $X_k + V_k - 1$ units still remain in the system. Summarizing we therefore have

$$X_{k+1} = \begin{cases} V_k & \text{if } X_k = 0, \\ X_k + V_k - 1, & \text{if } X_k > 0. \end{cases} \tag{3}$$

If the service of the $(k + 1)$th unit begins at time t and lasts for x unit times, then the arrivals from time t form, by (2), a *Poisson process* with parameter λ (rate of arrival), and the number of arrivals in the interval of length x is, according to Example (2.1.*a*), *Poisson-distributed* with parameter λx. If $G(x)$ is the distribution function of the serving time, we have thus

$$a_j = P[V_k = j] = \int_0^\infty \frac{(\lambda x)^j}{j!} e^{-\lambda x} dG(x). \tag{4}$$

The random variables $V_k, k = 1, 2, \ldots$, all have the same distribution (4). From (2) it follows further that the V_k are mutually *independent*.

Therefore, according to (3), X_{k+1} is only dependent on X_k but not on $X_{k-1}, X_{k-2}, \ldots$, i.e. for $X_k, k = 1, 2, \ldots$, we have the Markov property. $\{X_k, k = 1, 2, \ldots\}$ is an *embedded Markov chain* in the X_t process of the $M/G/1$-system. Its *transition matrix* is determined by the distribution $\{a_j, j = 0, 1, 2, \ldots\}$ of V_k as follows (see (3)):

$$\begin{bmatrix} a_0 & a_1 & a_2 & a_3 & a_4 & \cdots \\ a_0 & a_1 & a_2 & a_3 & a_4 & \cdots \\ 0 & a_0 & a_1 & a_2 & a_3 & \cdots \\ 0 & 0 & a_0 & a_1 & a_2 & \cdots \\ 0 & 0 & 0 & a_0 & a_1 & \cdots \\ \cdot & \cdot & \cdot & \cdot & \cdot & \\ \cdot & \cdot & \cdot & \cdot & \cdot & \\ \cdot & \cdot & \cdot & \cdot & \cdot & \end{bmatrix} \tag{5}$$

Examples

(*a*) *Constant serving times.* If the serving time is constant (deterministic) and equal to d (i.e. $G(x) = 0$ for $x < d$, $G(x) = 1$ for $x \geqslant d$, it follows from (4) that

$$a_j = \frac{(\lambda d)^j}{j!} e^{-\lambda d}.$$

This is a *Poisson distribution* (see Example (1.2.*d*)).

(*b*) *Exponentially distributed serving times.* If $G(x) = 1 - e^{-\mu x}$ the $M/G/1$-system reduces to the $M/M/1$-system as a special case. In this case we obtain

$$a_j = \int_0^\infty \frac{(\lambda x)^j}{j!} e^{-(\lambda + \mu)x} dx.$$

By partial integration we get $a_j = a_{j-1} \lambda/(\lambda + \mu)$ and thus $a_j = a_0 (\lambda/(\lambda + \mu))^j$. If we compute a_0 as well, we obtain

$$a_j = \frac{\mu}{\lambda + \mu} \left(\frac{\lambda}{\lambda + \mu} \right)^j,$$

a *geometric* distribution.   □

By (4), $a_j > 0$ for all $j = 0, 1, 2, \dots$. Since $a_0 > 0$ we have $0 \Leftrightarrow 0$ and $i \Rightarrow i - 1$ and thus $i \Rightarrow 0$. Since $a_i > 0$ for $i > 0$, we have $0 \Rightarrow i$ and therefore $i \Leftrightarrow 0$ and thus, finally, $i \Leftrightarrow j$ for every i and j. The embedded Markov chain is therefore *irreducible* and, because of $a_0 > 0$, also *aperiodic*.

Using Theorem 3.2.10 we shall first examine under which conditions the chain is *transient* or *recurrent*. The state 0 seems to suggest itself as a distinguished state for the application of Theorem 3.2.10. The equations (3.2.15) can be re-written as

$$a_1 y_1 + a_2 y_2 + a_3 y_3 + \dots = y_1,$$
$$a_1 y_2 + a_2 y_3 + \dots = y_2 - a_0 y_1,$$
$$a_1 y_3 + \dots = y_3 - a_0 y_2, \tag{6}$$
$$\vdots \qquad \vdots$$

In addition we shall consider the *inhomogeneous* system which we obtain by replacing the first equation of (6) by

$$a_1 y_1 + a_2 y_2 + a_3 y_3 + \ldots = y_1 - a_0 y_0 \tag{7}$$

with $y_0 = 1$. This inhomogeneous system has the solution $y_j = 1, j = 0, 1, 2, \ldots$, since the matrix (5) is *stochastic*.

There may be a second solution of the form $y_j = y^j$ with $y < 1, j = 0, 1, 2, \ldots$. If, in fact, $f(y) = a_1 y + a_2 y^2 + a_3 y^3 + \ldots$ and if we substitute $y_j = y^j$ into the inhomogeneous system, we obtain from every equation $f(y) = y - a_0$. The left hand side is *continuous* in the interval $[0, 1]$ and takes the values 0 and $1 - a_0$ at the boundaries of the interval. The right hand side takes the values $-a_0 < 0$ and $1 - a_0$ at the boundaries $y = 0$ and $y = 1$. Finally

$$f'(1) = \sum_{k=1}^{\infty} k a_k. \tag{8}$$

The equation $f(y) = y - a_0$ must thus have at least one more solution z, $0 < z < 1$, if $f'(1) > 1$ (see Fig. 1).

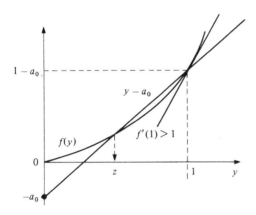

Fig. 1

The difference of two solutions of the inhomogeneous system is a solution of the homogeneous system (6). Therefore (6) has the *bounded* solution $y_i = 1 - z^i, i = 1, 2, \ldots$, if $f'(1) > 1$. By (8) and the definition of the probabilities $a_k, f'(1)$ is the *expected number of arrivals during one service*. Therefore, according to Theorem 3.2.10, the chain is *transient* if the *average number of arrivals during one service is greater than* 1. In this case there is in particular a positive probability that the system is never empty, i.e. that the chain will never again reach the state 0.

We might conjecture that the embedded Markov chain is positively recurrent if the average number of arrivals during one service (8) is less than 1.

In order to prove this we have to show according to Theorem 3.2.9 that there exists a *stationary distribution*.

The equations (3.2.10) for the embedded Markov chain read

$$p_j = a_j p_0 + \sum_{i=1}^{j+1} a_{j-i+1} p_i, \quad j = 0, 1, 2, \ldots . \tag{9}$$

If we introduce the quantities $r_k = 1 - a_0 - a_1 - \ldots - a_k, k = 0, 1, 2, \ldots,$ and sum the equations (9) successively we obtain the system

$$a_0 p_1 = r_0 p_0,$$

$$a_0 p_2 = r_1 p_0 + r_1 p_1, \tag{10}$$

$$a_0 p_3 = r_2 p_0 + r_2 p_1 + r_1 p_2,$$

$$\ldots \ldots,$$

from which $p_1, p_2, \ldots$ can be determined recursively as functions of p_0. This solution is unique for every value of p_0 and if $p_0 > 0$, then it follows also that $p_i \geqslant 0$ for $i = 1, 2, \ldots .$

If we sum the equations (10) and note that

$$r_0 + r_1 + r_2 + \ldots = \sum_{k=0}^{\infty} k a_k = r \tag{11}$$

(see Example (2.2.*a*)) then it follows that

$$(1 - r_0) \sum_{i=1}^{\infty} p_i = r p_0 + (r - r_0) \sum_{i=1}^{\infty} p_i$$

and from this, if $r \neq 1$, that

$$\sum_{i=1}^{\infty} p_i = \frac{r}{1-r} p_0. \tag{12}$$

If $r < 1$ and if we write $p_0 = 1 - r$, then $\{p_i, i = 0, 1, 2, \ldots\}$ becomes a distribution and, since this distribution is the solution of (9), it becomes the unique *stationary* distribution of the embedded Markov chain. If $r < 1$, i.e. if the *average number of arrivals during the duration of a service is less than* 1, the embedded Markov chain becomes *positively recurrent*; see Theorem 3.2.9.

Let $\{p_n, n = 0, 1, 2, \ldots\}$ be the stationary distribution of the embedded Markov chain which exists for $r < 1$, let $P(z)$ be the corresponding *generating function* (see (2.2.1)) and let $A(z)$ be the *generating function* of $\{a_n, n = 0, 1, 2, \ldots\}$, i.e. of the distribution of the number of arrivals during a service:

$$P(z) = \sum_{j=0}^{\infty} p_j z^j; \quad A(z) = \sum_{j=0}^{\infty} a_j z^j. \tag{13}$$

If we multiply the jth equation in (9) with z^j and sum over all j, it follows that

$$P(z) = p_0 \sum_{j=0}^{\infty} a_j z^j + \sum_{j=0}^{\infty} \sum_{i=1}^{j+1} a_{j-i+1} p_i z^j$$

$$= p_0 A(z) + z^{-1} \sum_{i=1}^{\infty} p_i z^i \sum_{j=i-1}^{\infty} a_{j-i+1} z^{j-i+1}$$

$$= p_0 A(z) + z^{-1} (P(z) - p_0) A(z) \tag{14}$$

and from this that

$$P(z) = \frac{(z-1)A(z)}{z - A(z)} p_0. \tag{15}$$

Formula (15) is usually written in a somewhat different form. It is advantageous at this point to introduce the *Laplace transform*,

$$g(s) = \int_0^{\infty} e^{-sx} dG(x), \quad s \geqslant 0, \tag{16}$$

of a distribution function $G(x)$ which is concentrated on the positive half-axis ($G(x) = 0$ for $x \leqslant 0$). The Laplace transform plays a similar role for non-discrete distributions as the generating function does for discrete distributions.

According to (13), (4) and (16) we have

$$A(z) = \sum_{j=0}^{\infty} z^j \int_0^{\infty} \frac{(\lambda x)^j}{j!} e^{-\lambda x} dG(x)$$

$$= \int_0^{\infty} e^{-\lambda x} \sum_{j=0}^{\infty} \frac{(\lambda z x)^j}{j!} dG(x) = \int_0^{\infty} e^{(\lambda z - \lambda)x} dG(x)$$

$$= g(\lambda - \lambda z), \tag{17}$$

where $g(s)$ is the *Laplace transform of the serving-time distribution* $G(x)$. If we substitute (17) together with $p_0 = 1 - r$ in (15) we obtain

$$P(z) = \frac{(z-1)g(\lambda - \lambda z)}{z - g(\lambda - \lambda z)} (1-r). \tag{18}$$

This is the famous *Pollaczek–Khinchin formula*.

Examples

(c) *Constant serving times*. If $G(x) = 0$ for $x < d$, and $G(x) = 1$ for $x \geqslant d$ (see Example (a)), then the Laplace transform of this degenerate distribution is equal to $g(x) = e^{-sd}$. For (18) it follows that

$$P(z) = \frac{(z-1)e^{-(\lambda - \lambda z)d}}{z - e^{-(\lambda - \lambda z)d}} (1-r).$$

For this function no simple series expansion can be given and thus the stationary distribution cannot be determined explicitly (a recursive computation from (10) is, however, possible). This is typical of many results in queueing theory: for the transforms (generating functions, Laplace transforms) there are many explicit formulae but only in special cases can explicit results be derived for the original quantities. Nevertheless such results can be useful for the computation of moments, as we shall see.   □

(d) *Exponentially distributed serving times (M/M/1-system).* The Laplace transform of the serving time distribution is in this case equal to

$$g(s) = \int_0^\infty e^{-sx} \mu e^{-\mu x} \, dx = \frac{\mu}{s + \mu}.$$

According to (17) we have

$$A(z) = \frac{\mu}{\lambda + \mu - \lambda z}$$

(for $|z| < 1$). By differentiating with respect to z we obtain $A'(z) = \lambda \mu / (\lambda + \mu - \lambda z)^2$ and thus, by Example (2.2.a), $r = A'(1) = \lambda / \mu$. Substituting this in (18) we get

$$P(z) = \frac{(1 - z)(1 - r)}{1 - rz + rz^2 - z} = \frac{1 - r}{1 - rz} = (1 - r) \sum_{j=0}^\infty (rz)^j.$$

We obtain therefore $p_j = (1 - r)r^j$, $r = \lambda / \mu$. By comparing this with Example (3.4.a) we can see that for the $M/M/1$-system the embedded discrete Markov chain has the *same* limit distribution as the continuous Markov chain of Example (3.4.a). This means that whether we observe the system at arbitrary times or only at exits of units, the distribution of the number of units in the system is (asymptotically) the same. *The same, by the way, is true for every M/G/1-system* and not just for the $M/M/1$-system, but this will be proved later (Example (7.3.j)).   □

From the Laplace transform of a distribution one can determine its moments relatively simply, just as the expected value of a discrete distribution can be determined easily from its generating function; cf. Example (2.2.a). Let X be a random variable with the distribution function $G(x)$, $G(x) = 0$ for $x \leqslant 0$. If we substitute in (16) the series expansion of the exponential function, we obtain

$$g(s) = \int_0^\infty \sum_{j=0}^\infty \frac{(-sx)^j}{j!} \, dG(x) = \sum_{j=0}^\infty (-1)^j E[X^j] \frac{s^j}{j!}.$$

The last expression, however, is the Taylor series expansion of $g(s)$ around $s = 0$. Therefore the jth derivative of g satisfies $g^{(j)}(0) = (-1)^j E[X^j]$. To illustrate this let us consider a Laplace transform of the exponential distribution with parameter μ, $g(s) = \mu/(s + \mu)$ (see Example (d)). From this we obtain the expected value $E[X] = -g^{(1)}(0) = 1/\mu$ and the variance $\text{Var}[X] = E[X^2] - (E[X])^2 = g^{(2)}(0) - (1/\mu)^2 = 2/\mu^2 - 1/\mu^2 = 1/\mu^2$ (see Example $(1.3.g)$).

The Pollaczek–Khinchin formula (18) can now be used for the computation of moments of the stationary distribution of the embedded Markov chains. The expected value is in particular determined by $P'(1)$. If we write (18) in the form

$$P(z)(z - g(\lambda - \lambda z)) = (z - 1)g(\lambda - \lambda z)(1 - r), \tag{19}$$

and if we differentiate (19) with respect to z and solve for $P'(z)$ the result is

$$P'(z) = \frac{g(\lambda - \lambda z) - (z - 1)\lambda g'(\lambda - \lambda z)}{z - g(\lambda - \lambda z)}(1 - r) - \frac{P(z)(1 + \lambda g'(\lambda - \lambda z))}{z - g(\lambda - \lambda z)}.$$

$$\tag{20}$$

Let $-g'(0) = \tau$ be the *expected value* and $g''(0) - (g'(0))^2 = \sigma^2$ be the *variance* of the serving time. If we let z tend to 1 in (20) and apply the rule of L'Hôpital, we obtain for the expected value of the stationary distribution of the embedded Markov chain

$$P'(1) = r + \frac{r^2(1 + \sigma^2/\tau^2)}{2(1 - r)}. \tag{21}$$

This is the *mean value formula by Pollaczek–Khinchin*. $P'(1)$ is the asymptotic average value of the number of units (queue length) which a departing unit leaves behind. As remarked in Example (d) $P'(1)$ is also the asymptotic expected value of the number of units at any given time. It is quite remarkable that these expected values depend for a given r only on the expected value and the variance of the serving time.

Example

(e) *Sensitivity of the $M/M/1$-system.* For the exponential distribution $\sigma^2 = \tau^2$. Therefore, according to (21), we have for the $M/M/1$-system $P'(1) = r/(1 - r)$; see also Example $(3.4.a)$. The ratio σ^2/τ^2 can be used as a measure for the serving time variations. The *smaller* (*larger*) these variations are, the *smaller* (*larger*) is the mean queue length $P'(1)$. $\qquad\square$

4.3. Networks of queueing systems

From the $M/M/s$-system, Examples $(3.3.c)$ and $(3.3.f)$, it became apparent that exponentially distributed inter-arrival times and serving times are particularly favourable for analysis since in this case the system can very easily be described by a Markov chain. It is thus only natural when trying to generalize the $M/M/s$-system to retain the possibility of using Markov chains.

Such generalizations are possible in several directions. In this section we introduce a class of more general systems which can not only be described by Markov chains but whose stationary distributions can, if they exist, also be derived explicitly! We come to such systems when considering *networks* of service areas which have to be passed through by units in a manner not yet specified. Every service area may be occupied by several servers as in the *M/M/s*-system with exponentially distributed serving times. It is obvious that such a network is of practical importance since queues tend not to arise in isolation!

First let us fix the assumptions. Let the network consist of $N \geqslant 1$ service areas which are numbered $n = 1, 2, \ldots, N$. We describe the state of the system at any time t by a vector $k = (k_1, k_2, \ldots, k_N)$, whose components k_i denote the number of units in the ith service area. $S(k) = k_1 + k_2 + \ldots + k_N$ is the total number of units in the system.

The rate of arrival of units to the system may now even depend on the total number of units in the system. We assume a function $\lambda(k)$, $k = 0, 1, 2, \ldots$, to be given in such a way that the probability of an arrival during the interval from t up to $t + h$ is equal to $h\lambda(S(k)) + o(h)$, if the state of the system at time t is equal to k. As usual $o(h)$ represents a quantity which tends to 0 faster than h for $h \to 0$ (see Section 3.3). In order to guarantee irreducibility for the Markov chain to be introduced later, let us assume that there exists an integer k_0, $0 < k_0 \leqslant \infty$, such that $\lambda(k) > 0$ for $k < k_0$ and $\lambda(k) = 0$ for $k \geqslant k_0$. If k_0 is finite we can, for example, simulate a case in which arriving units are rejected if k_0 or more units are in the system (see Example (*a*) below).

The serving rate of the service area n may depend on the service area itself as well as on the number of units already there. Assume that we are given a function $\mu(n, k)$ such that the probability of completing a service in the service area n during the interval from t up to $t + h$ is equal to $h\mu(n, k_n) + o(h)$ if there are k_n units in the service area n at time t. In particular we assume that $\mu(n, 0) = 0$ and $\mu(n, k) > 0$ for $k > 0$. If the nth service area consists, for instance, of m parallel servers as in the *M/M/s*-system, then $\mu(n, k) = \min(k, m)\mu$. Again we assume the probability of two or more arrivals and/or completions of service during the interval from t to $t + h$ to be of the order $o(h)$.

Let the order in which the service area in the network are passed through by a unit be determined by the probabilities r_{mn} of transition to the next area n after leaving the service area m. Let r_{0n} be the probability that a newly arrived unit goes to area n and let r_{mN+1} be the probability that the unit leaves the system after leaving the service area m. (r_{mn}), $m, n = 0, 1, 2, \ldots, N, N + 1$, is a *stochastic transition matrix* where $r_{m0} = 0$ for $m = 0, 1, \ldots, N, N + 1$; $r_{N+1 n} = 0$ for $n = 0, 1, 2, \ldots, N$; $r_{0 N+1} = 0$ and $r_{N+1 N+1} = 1$.

The order in which service areas are passed through by a unit is thus described by a discrete Markov chain with an absorbing state $N + 1$. In order that each unit should only visit a finite number of service areas, the states $n = 1, 2, \ldots, N$ have to be *transient*. If the $r_{0i}^{(n)}$ are the n-step transition probabilities from the state 0 then

$$y_i = \sum_{n=1}^{\infty} r_{0i}^{(n)}, \quad i = 1, 2, \ldots, N, \tag{1}$$

is the expected number of visits to a service area i before leaving the system. If the states i are transient, the series in (1) converges (see Theorem 3.2.2(c)) and the y_i are finite. Further to Theorem 3.2.10 we introduce at this point another *transience criterion.*

Theorem 1

If the states $i = 1, 2, \ldots, N$ are transient, then the y_i in (1) form the *uniquely* determined solution of the system of equations

$$y_i = r_{0i} + \sum_{j=1}^{N} y_j r_{ji}, \quad i = 1, 2, \ldots, N. \tag{2}$$

Conversely, if equations (2) have a *unique* solution, then the states $i = 1, 2, \ldots, N$ are *transient*.

Proof. Let the states $i = 1, 2, \ldots, N$ be transient so that the series (1) converge. If we sum the equations

$$r_{0i}^{(1)} = r_{0i},$$

$$r_{0i}^{(n)} = \sum_{j=1}^{N} r_{0j}^{(n-1)} r_{ji}, \quad n > 1, \tag{3}$$

(see (3.1.10)) over n, we realize that the y_i are solutions of (2).

If this system had a second solution the difference of the two solutions would have to satisfy the homogeneous system

$$z_i = \sum_{j=1}^{N} z_j r_{ji}, \quad i = 1, \ldots, N. \tag{4}$$

If we multiply the ith equation by r_{ih} and if we sum over i, we observe that (4) holds for $r_{ji}^{(2)}$ instead of r_{ji}. By induction on n it follows that

$$z_i = \sum_{j=1}^{N} z_j r_{ji}^{(n)}.$$

Since the states $i = 1, \ldots, N$ are transient we have $r_{ji}^{(n)} \to 0$ as $n \to \infty$ (Theorem 3.2.4) and hence $z_i = 0, i = 1, 2, \ldots, N$.

If on the other hand (2) has a solution and if some of the states

$i = 1, 2, \ldots, N$ are *positively recurrent* (there can be no null-recurrent states for a finite chain according to Theorem 3.2.7), there will exist states which form a closed class. The stationary distribution for this class forms a solution of (4) and the solution of (2) is thus not unique. The states $i = 1, 2, \ldots, N$ are therefore transient if and only if (2) has a unique solution. ∎

Examples

(*a*) *Modifications of M/M/s-systems.* The $M/M/s$-system as well as some of its modifications can be introduced within the framework of our assumptions. In these examples one always deals with a *single* service area with s servers so that $r_{01} = r_{12} = 1$ (2 = exit from the service area). First let us mention the modification of the *finite waiting room.* If there are more than $k \geqslant k_0$ units in the system, new arrivals have to be turned away. This can be done by setting $\lambda(k) = \lambda$ for $k < k_0$, $\lambda(k) = 0$ for $k \geqslant k_0$ where $k = k, S(k) = k$. If $k_0 = s$ we have a model for a bounded number of *parking spaces* where the parking spaces are considered as servers. Secondly, there is the case of a *finite number of units*, i.e. there are altogether only m units which can demand access to the system. If k units are in the system, then only $m - k$ are outside the system and are potential arrivals. In this case we write $\lambda(k) = (m - k)\lambda$ (see also Example (3.3.*d*)).   □

(*b*) *Self-service stores, multi-storey car parks, skiing resorts.* Here we show how to find models for some concrete systems within the framework of our present assumptions. The simple *self-service store* can be represented by two service areas in series: (1) the actual shop, and (2) the cash points. In this case $r_{01} = r_{12} = r_{23} = 1$ and all other $r_{ij} = 0$. For the service area 1 we can assume infinitely many servers (an unlimited number of people who could be in the shop), i.e. $\mu(1, k) = k\mu_1$. For s cash points the service area 2 is an s-server system with $\mu(2, k) = \min(k, s)\mu_2$. We can further assume $\lambda(S(k)) = \lambda$ for the arrivals. $1/\mu_1$ and $1/\mu_2$ are the average length of stay of customers in the store and the average service time respectively, where these times are assumed to be exponentially distributed.

For the *multi-storey car park* we may, for example, consider the following service areas: (1) entrance barrier, (2) parking area, (3) car wash, (4) exit barrier. The flow of units (cars) can be described by the r_{ij} as follows: $r_{01} = 1, r_{12} = 1, r_{23}, r_{24} > 0$ and $r_{23} + r_{24} = 1, r_{34} = 1, r_{45} = 1$, all other $r_{ij} = 0$. For one entrance barrier, one exit barrier and one car wash plant we may put $\mu(i, k) = \min(k, 1)\mu_i, i = 1, 3, 4$. Further for s parking lots in the parking area, we may put $\mu(2, k) = \min(k, s)\mu_2$. The $1/\mu_i$ are again the average serving times in the individual service areas. The serving times themselves have to be assumed as exponentially distributed in order to satisfy the assumptions. This shows the limitation of the model, since, for example, an exponential distribution of the serving time is hardly realistic for an automatic car wash plant. In spite of this the model can be useful. The limited car

Table 1

Transition	Probability
$k \to k(i +)$	$h\lambda(S(k))r_{0i} + o(h)$
$k \to k(i -)$	$h\mu(i, k_i)r_{iN+1} + o(h)$
$k \to k(i, j), i \neq j$	$h\mu(i, k_i)r_{ij} + o(h)$
$k \to k$	$1 - h\lambda(S(k)) - h \sum_{i=1}^{N} \mu(i, k_i)r_{iN+1} - h \sum_{i=1}^{N} \sum_{\substack{j=1 \\ j \neq i}}^{N} \mu(i, k_i)r_{ij} + o(h)$
	$= 1 - h\lambda(S(k)) - h \sum_{i=1}^{N} \mu(i, k_i)(1 - r_{ii}) + o(h)$
others	$o(h)$

parking capacity of the multi-storey car park can be taken into account by setting $\lambda(S(k)) = \lambda$ if $S(k) < s$, $\lambda(S(k)) = 0$ if $S(k) \geq s$.

In a *skiing resort* there are various means of transport available (ski-lifts, chair lifts, cable cars etc.) which are connected by skiing slopes. The means of transport and the slopes may be understood as service areas. The possible connections and flows of skiers between the service areas are to be represented by the r_{ij}. It is left to the reader to judge how realistic or unrealistic such a model would be for a skiing resort. □

After these preparations the *continuous Markov chain* can be introduced whose state at time t is defined by k. In order to describe the transitions of this chain let us introduce some notation: $k(i -) = (k_1, \ldots, k_{i-1}, k_i - 1, k_{i+1}, \ldots, k_N)$, $k(i +) = (k_1, \ldots, k_{i-1}, k_i + 1, k_{i+1}, \ldots, k_N)$ and $k(i, j) = (k_1, \ldots, k_{i-1}, k_i - 1, k_{i+1}, \ldots, k_{j-1}, k_j + 1, k_{j+1}, \ldots, k_N)$, where $k(i, i) = k$. From the assumptions introduced above we obtain the transition probabilities for the interval from t to $t + h$ given in Table 1.

As in Section 3.3 (especially Examples (3.3.c) and (3.3.d)) we can deduce from this the elements of the differential matrix of the Markov chain. The elements of the differential matrix are $q(k, k(i +)) = \lambda(S(k))r_{0i}$, $q(k, k(i -)) = \mu(i, k_i)r_{iN+1}$, $q(k, k(i, j)) = \mu(i, k_i)r_{ij}$, $i \neq j$ and $q(k) = \lambda S(k) + \Sigma_{i=1}^{N} \mu(i, k_i)(1 - r_{ii})$.

All other elements of the differential matrix vanish. If there exists a stationary distribution $p(k)$ for the Markov chain, by (3.4.2), it has to satisfy the equations

$$\left\{ \lambda S(k) + \sum_{i=1}^{N} \mu(i, k_i)(1 - r_{ii}) \right\} p(k) = \sum_{i=1}^{N} p(k(i -))\lambda(S(k) - 1)r_{0i}$$

$$+ \sum_{i=1}^{N} p(k(i +))\mu(i, k_i + 1)r_{iN+1} + \sum_{i=1}^{N} \sum_{\substack{j=1 \\ j \neq i}}^{N} p(k(i, j))\mu(j, k_j + 1)r_{ji}.$$
$$(5)$$

To these equations we have to add the *boundary conditions* $p(k) = 0$ for vectors k with at least one *negative* component. We then have in particular $p(k(i-)) = 0, p(k(i, j)) = 0$ for states k with $k_i = 0$.

In the following theorem we shall give a solution of these equations in *explicit* form. At first the reader should verify that the chain is *irreducible* if we restrict ourselves to states k with $S(k) \leqslant k_0$. According to Theorem 3.4.1 this solution must be the *uniquely* determined stationary distribution of the Markov chain.

Theorem 2
Let

$$G(k) = \prod_{i=0}^{k-1} \lambda(i), \quad k = 0, 1, 2, \ldots, \tag{6}$$

$$F(k) = \prod_{n=1}^{N} \prod_{i=1}^{k_n} \frac{y_n}{\mu(n, i)}, \quad k = (k_1, \ldots, k_N), \tag{7}$$

where empty products have the value 1 and where $y_n, n = 1, 2, \ldots, N$, is the assumed unique solution of (2). If

$$\frac{1}{p} = \sum_{k=0}^{\infty} G(k) \sum_{k:S(k)=k} F(k) < \infty, \tag{8}$$

then

$$p(k) = pG(S(k))F(k) \tag{9}$$

is a distribution and a solution of (5), i.e. the stationary distribution of the corresponding Markov chain.

Proof. Equation (8) is the normalization of (9) to a distribution. The normalization factor p cancels in (5) and it thus suffices to consider $G(S(k))F(k)$. It follows from definition (6) that

$$G(S(k(i-))) = G(S(k) - 1) = G(S(k))/\lambda(S(k) - 1),$$

$$G(S(k(i+))) = G(S(k) + 1) = G(S(k))\lambda(S(k)),$$

$$G(S(k(i, j))) = G(S(k)),$$

and from definition (7) that

$$F(k(i-)) = \frac{F(k)}{y_i/\mu(i, k_i)},$$

$$F(k(i+)) = F(k)\frac{y_i}{\mu(i, k_i + 1)},$$

$$F(k(i, j)) = F(k)\frac{y_j/\mu(j, k_j + 1)}{y_i/\mu(i, k_i)}.$$

Since $\mu(i, 0) = 0$ it follows that $F(k(i-)) = 0$ and $F(k(i, j)) = 0$ for states k with $k_i = 0$. Hence, (9) satisfies the boundary condition of the system (5).

To simplify (5) it is recommended to move the sum of the terms $\mu(i, k_i)r_{ii}$ to the right hand side. If we then substitute (9) on the right hand side of (5) it follows, after several transformations and by considering (2), that the right hand side of (5) is equal to

$$p(k)\left\{\sum_{i=1}^{N} \mu(i, k_i) + \lambda(S(k)) \sum_{i=1}^{N} y_i r_{iN+1}\right\}. \tag{10}$$

If we sum the equations (2) over $i = 1, \ldots, N$ and note that the matrix of the r_{ji} is stochastic we realize that the sum of the $y_i r_{iN+1}$ in (10) must be equal to 1. Therefore (10) is equal to the modified left hand side of (5). This proves that (9) is a solution of (5). ∎

Examples

(c) *Networks of M/M/s-systems.* If we assume a *constant* rate of arrival $\lambda(k) = \lambda$ for $k = 0, 1, 2, \ldots, G(k)$ becomes equal to λ^k. If we define

$$w_n(k) = \prod_{i=1}^{k} \frac{\lambda y_n}{\mu(n, i)}, \quad p_n(k) = w_n(k)\bigg/\sum_{i=0}^{\infty} w_n(i),$$

then $\{p_n(k), k = 0, 1, 2, \ldots\}$ forms a *probability distribution*. Furthermore we have

$$F(k) = \prod_{n=1}^{N} \frac{w_n(k_n)}{\lambda^{k_n}}$$

(see (7)) and it therefore follows by Theorem 2 for the stationary distribution for a constant rate of arrival that

$$p(k) = p \prod_{n=1}^{N} w_n(k_n) = \prod_{n=1}^{N} p_n(k_n),$$

since the p_n form probability distributions for every $n = 1, 2, \ldots, N$. It can be seen from this that the p_n are the *marginal distributions* of the multidimensional distribution $p(k)$. Thus $\{p_n(k), k = 0, 1, 2, \ldots\}$ is the distribution of the number of units in the service area n. The limit distributions of the number of units in the service areas are, for $t \to \infty$, mutually *independent* since $p(k)$ is the product of the $p_n(k_n)$! This result is remarkable.

If each of the N service areas consists of s_n parallel servers with exponentially distributed serving times as in the $M/M/s$-systems and if the serving rate of the servers in the service area n is equal to μ_n, then $\mu(n, i) = \min(i, s_n)\mu_n$. Thus $w_n(0) = 1$ and

$$
p_n(k) = p_n(0)w_n(k) =
\begin{cases}
\dfrac{p_n(0)(\lambda y_n)^k}{k!\,\mu_n^k}, & k \leqslant s_n, \\[3mm]
\dfrac{p_n(0)(\lambda y_n)^k}{s_n!\,s_n^{k-s}\mu_n^k}, & k > s_n.
\end{cases}
$$

A comparison of this result with Example $(3.4.a)$ shows that $p_n(k)$ is nothing but the distribution of the number of units in an $M/M/s_n$-system with *arrival rate* λy_n and service rate μ_n for the servers. The network can therefore be decomposed into $M/M/s_n$-systems with arrival rates λy_n which can be treated *independently* of each other. We have only first to determine the average number of visitors y_n to the service areas $n = 1, 2, \ldots, N$ from equations (2).

For a self-service store, for instance, (Example (b)) the distribution of customers shopping in the store is thus equal to that of an $M/M/\infty$-system with arrival rate λ and service rate μ_1 and the distribution of customers at the s cash points is equal to that of an $M/M/s$-system with arrival rate λ and service rate μ_2, and both distributions are mutually independent. $\square$

(d) *Closed networks.* As before we assume N service areas and transition probabilities $r_{ij}, i, j = 1, 2, \ldots, N$, for the transitions of units from one service area to the next. Let there be a total of M units in the system and let neither arrivals nor departures be permitted so that the same M units circulate in the system for ever. The $r_{ij}, i, j = 1, 2, \ldots, N$ should define an irreducible Markov chain. Such a system is called *closed*; it does not satisfy *a priori* the conditions of Theorem 2 since, for a finite irreducible chain, all states have to be recurrent (Theorem 3.2.7); according to Theorem 1, however, the system (2) no longer has a uniquely determined solution. By isolating a service area — let us say area 1 — and by considering the *remaining* network of the service areas $n = 2, \ldots, N$ we can, nevertheless, apply Theorem 2. If $k' = (k_2, \ldots, k_n)$ is the state vector for the reduced network, then obviously the arrival rates for the reduced networks are $\lambda(S(k')) = (1 - r_{11})\mu(1, M - S(k'))$ for $S(k') < M$, $\lambda(S(k')) = 0$ for $S(k') = M$. The equations (2) for the reduced network take on the form

$$
y_i = \frac{r_{1i}}{1 - r_{11}} + \sum_{j=2}^{N} y_j r_{ji}, \quad i = 2, \ldots, N,
$$

since the probability of a transition from 1 to i has obviously to be replaced by the conditional probability, given a jump from 1 to a state distinct from 1! Since state 1 can be reached with probability 1 from every state $i = 2, \ldots, N$

(Theorem 3.2.8), the states $i = 2, \ldots, N$ become transient if state 1 is made absorbing. Therefore the above system of equations has, by Theorem 1, a unique solution.

If we apply Theorem 2 and put $y_1 = 1/(1 - r_{11})$, $k_1 = M - S(k')$, $k = (k_1, k_2, \ldots, k_N)$, where $k_1 + k_2 + \ldots + k_{N_r} = M$, we find that

$$G(S(k')) = \prod_{i=0}^{S(k')-1} \frac{\mu(1, M - i)}{y_1} = \eta \prod_{i=1}^{k_1} \frac{y_1}{\mu(1, i)},$$

where $\eta = (\mu(1, 1)\mu(1, 2) \ldots \mu(1, M))/y_1^M$, is independent of k. Thus it follows for the stationary boundary distribution $p(k)$ from (9) that

$$p(k) = p \prod_{n=1}^{N} w_n(k_n), \quad w_n(k_n) = \prod_{i=1}^{k_n} \frac{y_n}{\mu(n, i)}, \quad \begin{cases} n = 1, 2, \ldots, N, \\ k_1 + k_2 + \ldots + k_N = M, \end{cases}$$

where p is the normalizing constant in which the above factor η is absorbed. $y_1 = 1/(1 - r_{11})$ and $y_2, \ldots, y_N$ satisfy the equations

$$y_i = \sum_{j=1}^{N} y_j r_{ji}, \quad i = 1, 2, \ldots, N.$$

If r_{ij} determines an irreducible and aperiodic Markov chain, then it also has a stationary distribution $x_1, x_2, \ldots, x_N$ which is again a solution of these equations and we have $(y_1, y_2, \ldots, y_N) = \rho(x_1, x_2, \ldots, x_N)$. This is why in $w_n(k_n)$ above we may replace the quantity y_n by x_n, where in $p(k)$ only the normalizing constant p has to be replaced by $p\rho^M$. $\qquad\square$

(e) *Closed networks of M/M/s-systems.* If Example (d) is specialized in such a way that $\mu(n, i) = \min(i, s_n)\mu_n$ (s_n servers at service point n), then

$$w_n(k) = \begin{cases} \dfrac{x_n^k}{k! \, \mu_n^k}, & k \leqslant s_n, \\[3ex] \dfrac{x_n^k}{s_n! \, s_n^{k-s_n} \mu_n^k}, & k > s_n. \end{cases}$$

These are again the formulae for $M/M/s_n$-systems with arrival rates x_n, where the x_n are determined from the equations of the last example. Although $p(k)$ is essentially again the product of the distributions of the individual service areas, as in Example (c), we cannot say that these distributions are independent since they are connected by the side condition $k_1 + k_2 + \ldots + k_N = M$.

The machine maintenance problem of Example (3.4.b) can be represented as a closed network with two service areas. Service area 1 consists of m servers for $M = m$ machines, service area 2 of n servers, corresponding to n mechanics.

Furthermore $r_{12} = r_{21} = 1$ and $r_{11} = r_{22} = 0$. This Markov chain is not aperiodic, but periodic with period 2. Nevertheless there exists a distribution which satisfies the equations of the y_i, $i = 1, 2$, in the last example ($x_1 = x_2 = \frac{1}{2}$) and the derived solution formulae can be applied with these values. In the notation of Example (3.4.b), $\mu_1 = \lambda$ and $\mu_2 = \mu$. $\qquad\qquad\square$

Notes to Chapter 4

There is an enormous amount of literature on queueing systems. A lot of very special models have been proposed and analysed. Nevertheless a number of quite general results have emerged as, for example, in the domain of networks of queueing systems; see [8], [9], [10] and especially [21] for more recent developments in this area. Also quite a number of different methodological approaches have been developed, the method of embedded Markov chains being only one among many. This method is due to *Kendall* [13]. Older textbooks such as [5], [12], [18], [19], [24] have a tendency to stress one particular method. Modern textbooks and monographs on the other hand try to find a synthesis, a unification of the theory of queueing systems; see [2], [4], [6], [7], [22]. For the *GI*/1/1-system especially there have been quite a few achievements in this direction. [1] gives a survey of the developments in queueing theory up to 1968.

There are some important questions and problems which have not been discussed here. Among others we should mention

(a) the statistical analysis of queueing systems (e.g. parameter estimation for queueing models),

(b) the optimal design and control of queueing systems,

(c) the numerical analysis of queueing models.

In [3] the reader will find some contributions to these questions. Concerning optimization it may be mentioned that methods introduced in the next chapter can be applied. For the numerical analysis of queueing systems we refer also to [20] and [21]. In this context approximations (see [3]) and bounds (see [15]) are of interest. But, especially for very complex systems, simulation and Monte-Carlo methods will be applied (see Chapter 6).

For practical applications of queueing theory we refer to [17]. Important applications arise in the design and analysis of computer and communication systems; see [16] and also [3.2]. The classical domain of application is in telephone systems; [23].

Queueing systems are also important in scheduling problems where priority rules for the scheduling of jobs have to be studied; see [11]. Other queueing problems arise in traffic systems (road, rail, air and water), in transportation and logistics, inventory and maintenance.

5
Dynamical optimization

5.1. Sequential decision problems

In this chapter we shall consider stochastic processes, in particular
Markov chains, whose evolutions can be influenced and controlled by a
sequence of *decisions*. In doing so we try to determine which decisions to
choose so that the evolution of the process is controlled *optimally* with
respect to an *a priori* criterion. In this section we introduce a finite-step
stochastic decision or optimization problem of a general form. Afterwards we
specialize the problem and investigate a limiting process which leads to an
infinite-step decision problem. This will serve as a transition to infinite-step
Markovian decision problems which will be considered in the two following
sections.

Let a system have the states $i \in I = \{1, 2, \ldots, N\}$. Assume that a sequence
of decisions has to be made, where the state of the system is known before
each one of these decisions. The index of the decision sequence $n = 0, 1,$
$2, \ldots$ can be interpreted as *time*. If the system is in state i, an *action a* is to
be chosen from a finite *set of actions* A_i. Let $A = A_1 \cup \ldots \cup A_N$. The *past* of
the system at time n is to be denoted by $y_n = \{i_0, a_0, i_1, a_1, \ldots, i_{n-1}, a_{n-1}, i_n\}$,
i.e. by the list of the states already passed through and by the chosen actions.
Let $A(y_n)$ denote the set A_i if i is the *last* state in y_n. Let $H_0 = I$ and
$H_{n+1} = \{(y_n, a, i): y_n \in H_n, a \in A(y_n), i \in I\}$; H_n are the sets of the *admissible*
pasts at times $n = 0, 1, 2, \ldots$.

Let us furthermore assume a probability distribution $p(i), i \in I$, of the
initial state at $n = 0$. Moreover, let there be given *transition probabilities*
$p_n(y_n, a, i)$ which determine the probability of reaching the state i at time
$n + 1$ if at time n we have the past y_n and if we have chosen the action
$a \in A(y_n)$. Finally let $r_n(y_n, a)$ denote a *pay-off* at the $(n + 1)$th decision, if
at time n we have the past y_n and the action $a \in A(y_n)$ has been chosen.

Let $f_n: H_n \to A$ be maps which assign to each admissible past y_n an admissible action $f_n(y_n) \in A(y_n) \subset A$. A sequence $f = \{f_0, f_1, f_2, \ldots\}$ of such maps
is a *policy*.

Every policy f determines a *stochastic process* as follows: an initial state i_0
is determined randomly by the distribution $p(i)$. Then we choose the action
$f_0(i_0)$. After this the next state i_1 is reached randomly, according to the

transition probabilities $p_0(i_0, f_0(i_0), i_1)$, whereupon the action $f_1(i_0, f_0(i_0), i_1)$ is chosen, the system changes over into the next state i_2 with the transition probability $p_1(i_0, f_0(i_0), i_1, f_1(i_0, f_0(i_0), i_1), i_2)$, etc. At the same time a sequence of pay-offs $r_0(i_0, f_0(i_0)), r_1(i_0, f_0(i_0), i_1, f_1(i_0), i_0))$, etc. is determined.

With every policy f one obtains in particular the expected values of the pay-offs determined by it. Let $F_{mf}^h(y_{h-m})$ be the expected value of the sum of the pay-offs under the policy f over the periods $n = h - m, \ldots, h$, if at time $h - m$ we have the past y_{h-m}. Obviously we have the following recursive relations

$$F_{0f}^h(y_h) = r_h(y_h, f_h(y_h)), \tag{1}$$

$$F_{mf}^h(y_{h-m}) = r_{h-m}(y_{h-m}, f_{h-m}(y_{h-m}))$$

$$+ \sum_{i=1}^{N} p_{h-m}(y_{h-m}, f_{h-m}(y_{h-m}), i) F_{m-1f}^h(y_{h-m}, f_{h-m}(y_{h-m}), i),$$

$$m = 1, 2, \ldots, h. \tag{2}$$

(2) is explained as follows: for the past y_{h-m} we choose the action $f_{h-m}(y_{h-m})$. Connected with this is the pay-off $r_{h-m}(y_{h-m}, f_{h-m}(y_{h-m}))$ for the period $h - m$. At the same time the system changes with probability $p_{h-m}(y_{h-m}, f_{h-m}(y_{h-m}), i)$ over into the state i at time $h - m + 1$. The expected pay-off for the remaining periods $h - m + 1, \ldots, h$ is then equal to $F_{m-1f}^h(y_{h-m}, f_{h-m}(y_{h-m}), i)$.

The expected value G_{hf} of the total pay-off during the first $h + 1$ periods $n = 0, 1, \ldots, h$ is thus obtained as

$$G_{hf} = \sum_{i=1}^{N} p(i) F_{hf}^h(i). \tag{3}$$

After these preparations the following *h-step stochastic sequential decision problem* can be formulated: *for a given planning period $h \geq 0$ one should choose a policy f^h which maximizes G_{hf}.* Period h is also called the *planning horizon.* A policy f^h which maximizes G_{hf} is called *optimal for the planning horizon h*, or simply optimal, if the planning horizon is kept fixed. G_{hf} only depends on $f_0, f_1, \ldots, f_h$. It suffices therefore to determine the first $h + 1$ elements of f^h; $f_{h+1}^h, f_{h+2}^h, \ldots$ are irrelevant for this problem.

The problem can be solved by complete enumeration, by computing G_{hf} for all possible policies f which differ in the first $h + 1$ terms $f_0, \ldots, f_h$. This is possible since there are only finitely many such policies. The so-called *optimality principle* offers a more efficient way of solving this problem and its basic idea can be formulated as follows: for every past y_{h-m} at time $h - m$ the remaining decisions for $n = h - m, \ldots, h$ are to be chosen *optimally*, i.e. in such a way that the expected value of the pay-offs during the remaining periods, given the past y_{h-m}, becomes maximal. On the basis of this idea we look for a policy g for which

$$F_{mg}^h(y_{h-m}) \geqslant F_{mf}^h(y_{h-m}) \tag{4}$$

for all $m = 0, 1, 2, \ldots$, for all $y_{h-m} \in H_{h-m}$ and *for all policies f.* For such a policy g, (4) holds in particular for $m = h$ and therefore this policy is certainly optimal by (3).

Allowing ourselves to be guided by this idea, we define

$$V_m^h(y_{h-m}) = \max_f F_{mf}^h(y_{h-m}) \tag{5}$$

as the maximal obtainable expected pay-off over the periods $n = h - m, \ldots, h$ if at $h - m$ we have the past $y_{h-m} \in H_{h-m}$. Note that

$$V_0^h(y_h) = \max_{a \in A(y_h)} \{r_h(y_h, a)\}, \quad y_h \in H_h. \tag{6}$$

The following theorem now gives the key for applying the optimum principle for the determination of an optimal policy $g = f^h$.

Theorem 1
For $m = 1, 2, \ldots, h$ and for all $y_{h-m} \in H_{h-m}$ we have

$$V_m^h(y_{h-m}) = \max_{a \in A(y_{h-m})} \{r_{h-m}(y_{h-m}, a)$$
$$+ \sum_{i=1}^N p_{h-m}(y_{h-m}, a, i) V_{m-1}^h(y_{h-m}, a, i)\}. \tag{7}$$

Proof. Let $V_m^h(y_{h-m})$ for $m = 0$ be defined by (6) and for $m = 1, 2, \ldots, h$ by (7). First we show by induction on m that

$$V_m^h(y_{h-m}) \geqslant F_{mf}^h(y_{h-m}) \tag{8}$$

for all policies f, all $m = 0, 1, 2, \ldots, h$ and all pasts $y_{h-m} \in H_{h-m}$. For $m = 0$ (8) is implied by (6). Suppose (8) holds for $m - 1$. It then follows from (2) and (7) that

$$F_{mf}^h(y_{h-m}) \leqslant r_{h-m}(y_{h-m}, f_{h-m}(y_{h-m}))$$
$$^. + \sum_{i=1}^N p_{h-m}(y_{h-m}, f_{h-m}(y_{h-m}), i) V_{m-1}^h(y_{h-m}, f_{h-m}(y_{h-m}), i)$$
$$\leqslant \max_{a \in A(y_{h-m})} \{r_{h-m}(y_{h-m}, a)$$
$$+ \sum_{i=1}^N p_{h-m}(y_{h-m}, a, i) V_{m-1}^h(y_{h-m}, a, i)\}$$
$$= V_m^h(y_{h-m}).$$

On the other hand if we take for every $y_{h-m} \in H_{h-m}$ an action $a \in A(y_{h-m})$ which *maximizes* (7), put $g_{h-m}(y_{h-m}) = a$ and if we define a policy g whose first $h + 1$ elements are equal to $g_0, g_1, \ldots, g_h$, then it follows from (1), (2), (6), (7) that

$$F_{mg}^h(y_{h-m}) = V_m^h(y_{h-m}). \tag{9}$$

According to (8) and (9), $V_m^h(y_{h-m})$ satisfies (5) and we have proved the theorem.   ∎

This proof contains in its second part a method for determining an optimal policy. The policy g constructed there obviously satisfies (4) because of (8) and (9), and is therefore optimal. Concretely one proceeds as follows:

1. For all $y_h \in H_h$ we maximize (6). If the maximum in (6) for a y_h is reached by an action a, we put $g_h(y_h) = a$. This determines at the same time $V_0^h(y_h) = F_{0g}^h(y_h)$.

2. For $m = 1, 2, \ldots, h$ and all y_{h-m} one maximizes (7) successively. If the maximum in (7) is achieved by an action a for a given m and y_{h-m} we put $g_{h-m}(y_{h-m}) = a$. At the same time we have determined $V_m^h(y_{h-m}) = F_{mg}^h(y_{h-m})$.

This process is called *backward recursion of dynamical optimization*. It leads to an optimal policy, a fact which we shall express in the following theorem for reasons of completeness.

Theorem 2

Every policy g which has been obtained by backward recursion of dynamical optimization is optimal, i.e. $G_{hf} \leqslant G_{hg}$ holds for every policy f.

Proof. The proof is already contained in the above remark. Suppose the policy g satisfies (4), (9). From (3) it follows for every policy f that

$$G_{hf} = \sum_{i=1}^N p(i)F_{hf}^h(i) \leqslant \sum_{i=1}^N p(i)V_h^h(i) = \sum_{i=1}^N p(i)F_{hg}^h(i) = G_{hg}. \quad ∎$$

We would like to emphasize that, as a consequence of these results the optimal policy is independent of the initial distribution $p(i)$.

The computation of an optimal policy by backward recursion can be very complicated in general since one has to compute g_{h-m} and V_m^h for all $y_{h-m} \in H_{h-m}, m = 0, 1, 2, \ldots, h$. Although the number of admissible pasts y_{h-m} is finite, it might be very large, in particular for large h and small m. In many cases the transition probabilities p_n and the pay-offs r_n are simple functions $z_n(y_n)$ of the past $y_n, p_n(y_n, a, i) = p_n(z_n(y_n), a, i), r_n(y_n, a) = r_n(z_n(y_n), a)$. In this case the computation can be considerably simplified since g_{h-m} and V_m^h have only to be determined for the possible values of z_n as can be seen from the formulae (6) and (7) of the backward recursion.

Such a special case occurs if p_n and r_n depend only on the *last* state of the past y_n. If $y_n = \{\ldots, i_{n-1}, a_{n-1}, i\}$, then we have $p_n(y_n, a, j) = p(i, a, j) = p_{ij}^a$, $r_n(y_n, a) = r_n(i, a) = r_{ni}^a$, where we assume additionally that the transition probabilities do not depend on time n, i.e. are *homogeneous* in time. Such a special decision problem is called *Markovian*.

From the backward recursion it follows by induction on m that $g_{h-m}(y_{h-m}) = g_{h-m}(i_{h-m})$ and $V_m^h(y_{h-m}) = V_m^h(i_{h-m})$, if i_{h-m} is the last state of y_{h-m}. The optimal decision for every past depends only on the *last* state. The same is true for the maximum pay-off for the remaining periods. This means that for every $m = 0, 1, \ldots, h$ the g_{h-m} and V_m^h only have to be computed for the N states $i = 1, 2, \ldots, N$ which means a substantial simplification of the computation! A policy f in which f_n depends only on the last state of the past y_n, is called *Markovian*. For Markovian decision problems there thus exists a Markovian decision policy which is optimal.

If a Markovian policy is given and if the system at time n is in state i, then the required decision is given by $f_n(i)$ and the transition probabilities into the next state $p(i, f_n(i), j)$ are *independent* of the past. Therefore a Markovian policy determines for a Markovian decision problem in general an *inhomogeneous discrete Markov chain*.

Examples

(a) *Maintenance.* Suppose that a system, a machine, a plant or an installation is subjected to constant wear and tear and is inspected at regular intervals. At an inspection the state of the system may be classified completely by one of the states $i = 0, 1, 2, \ldots, N$. Here state 0 signifies no wear and tear at all, and N signifies complete deterioration; the states in between indicate varying degrees of deterioration. Let the process of deterioration between the inspections be Markovian with transition probabilities p_{ij}. The fact that wear and tear can only increase if no appropriate measures are taken, can be expressed by the conditions $p_{ij} = 0$ for $j < i$, so that in particular $p_{NN} = 1$.

At each inspection, however, one can decide maintenance of the system which returns the system to state 0. We therefore have sets of action $A_i = \{0, 1\} = A$ where 0 means no maintenance and 1 indicates maintenance. We thus have $p_{ij}^0 = p_{ij}, p_{ij}^1 = p_{0j}$. It can further be assumed that the running cost per time interval c_i of the system depends on the state of wear i. Furthermore there is the cost of maintenance d_i so that $r_i^0 = c_i$ and $r_i^1 = c_0 + d_i$. In this example r does not depend on the time n.

The problem is finding a maintenance policy which *minimizes* the expected total cost over h inspection periods. This is a Markovian decision problem although in the form of minimization instead of maximization. We can maximize $-r_i^a$ or, equivalently, replace max by min in (6) and (7).

The equations (7) of the backward recursion in this example have the form

$$V_m^h(i) = \min \left\{ c_i + \sum_{i=1}^{N} p_{ij} V_{m-1}^h(j); c_0 + d_i + \sum_{j=0}^{N} p_{0j} V_{m-1}^h(j) \right\}$$

and $g_{h-m}(i) = 0$ or 1 depending on which of the two terms in the above minimization takes on the smaller value.   □

(*b*) *Replacement problems* (see also Example (3.1.*c*)). If one owns a machine of age i, one has to decide whether to keep it or replace it. At a maximal age N the machine is assumed to be worthless and has to be replaced in any case. Let the probability that a machine of age i survives another period and reaches age $i + 1$ (i.e. is not replaced) be equal to p_i. If it does not survive, its age is put equal to N. For $i = 0, 1, \ldots, N - 1$ we have the sets of action $A_i = \{n, k: k = 0, 1, \ldots, N - 1\}$ where n means no replacement while k means replacement by a machine of age k. We assume therefore that there exists a second-hand market (for instance for cars) where used machines can be purchased. Furthermore $A_N = \{k: k = 0, 1, \ldots, N - 1\}$.

Thus $p_{ii+1}^n = p_i, p_{iN}^n = 1 - p_i, p_{ij}^n = 0$ for $j \neq i + 1, N$ and $p_{ik+1}^k = p_k$, $p_{iN}^k = 1 - p_k$ and $p_{ij}^k = 0$ for $j \neq k + 1, N$. For the pay-offs the following data are relevant:

C_i purchasing price of a machine of age i

T_i selling price of a machine of age i

B_i expected running cost of one period for a machine of age i.

The pay-offs are then defined as follows: $r_i^n = B_i, r_i^k = C_k - T_i + B_k$. A replacement policy has to be determined which minimizes the expected cost over a planning horizon h.

Again we deal with a Markovian decision problem and the backward recursion is based on

$$V_m^h(i) = \min\Big\{ B_i + p_i V_{m-1}^h(i + 1) + (1 - p_i) V_{m-1}^h(N);$$

$$\min_{k=0,\ldots,N-1} [C_k - T_i + B_k + p_k V_{m-1}^h(k + 1) + (1 - p_k) V_{m-1}^h(N)] \Big\},$$

$$i = 0, 1, \ldots, N - 1,$$

$$V_m^h(N) = \min_{k=0,\ldots,N-1} \{C_k - T_N + B_k + p_k V_{m-1}^h(k + 1) + (1 - p_k) V_{m-1}^h(N)\}.$$

                                                                □

The pay-offs r_{ni}^a often take on the form $\beta^n r_i^a$, where β is a *discount factor*. This is the case of the two examples above, where we have $\beta = 1$, but which also make sense for $\beta < 1$. The $\beta^n r_i^a$ are the pay-offs r_i^a at the nth period, discounted for time 0. Then $V_m^h(i)$ is the expected value of the pay-offs of the periods $n = h - m, \ldots, h$, discounted for time 0, and $v_m^h(i) = \beta^{-(h-m)} V_m^h(i)$ are the expected values of the same pay-offs but discounted for time $h - m$. According to (6) and (7) we have

$$v_0^h(i) = \max_{a \in A_i} \{r_i^a\},$$

$$v_m^h(i) = \max_{a \in A_i} \left(r_i^a + \beta \sum_{j=1}^{N} p_{ij}^a v_{m-1}^h(j) \right), \quad m = 1, \ldots, h. \tag{10}$$

(10) implies that $v_m^h(i)$ does not depend on h, $v_m^h(i) = v_m^k(i)$, *but only on the number of periods m up to the end of the planning horizon.* For this reason we shall write in future $v_m^h(i) = v_m(i)$.

Examples

(c) *A computational example.* Let us consider an automatic production process in which certain parameters have to be adjusted very carefully so that the output is of satisfactory quality (state 1). If we let the process run without adjustment the parameters can easily change so that the quality of the output decreases and is only partly satisfactory (state 2). The process is now examined periodically for its state. After every inspection one can decide whether to adjust the parameters (action 1) or not (action 2).

Let us assume the following transition probabilities between the states over one period:

$$p_{11}^1 = 0.9, \quad p_{12}^1 = 0.1; \quad p_{11}^2 = 0.6, \quad p_{12}^2 = 0.4,$$
$$p_{21}^1 = 0.7, \quad p_{22}^1 = 0.3; \quad p_{21}^2 = 0.2, \quad p_{22}^2 = 0.8.$$

Assume furthermore that the following costs are connected with various states and actions (for unsatisfactory quality and re-adjustment)

$$r_1^1 = 5, \quad r_1^2 = 0; \quad r_2^1 = 20, \quad r_2^2 = 15.$$

We have to determine an optimal policy of parameter re-adjustment which minimizes the expected total cost over m periods. This is a Markovian decision problem which can be solved by backward recursion according to (10). The discount factor is chosen as $\beta = 0.9$. For the last period ($m = 0$) we find

$$v_0(1) = \min\{5, 0\} = 0; \qquad g_0(1) = 2,$$
$$v_0(2) = \min\{20, 15\} = 15; \qquad g_0(2) = 2.$$

For the last two periods we obtain

$$v_1(1) = \min \{5 + 0.9(0\ 9 \times 0 + 0.1 \times 15); 0.9(0.6 \times 0 + 0.4 \times 15)\}$$
$$= \min \{6.35; 5.4\} = 5.4, \quad g_1(1) = 2;$$
$$v_1(2) = \min \{20 + 0.9(0.7 \times 0 + 0.3 \times 15); 15 + 0.9(0.2 \times 0 + 0.8 \times 15)\}$$
$$= \min \{24.05; 25.8\} = 24.05, \quad g_1(2) = 1.$$

In this way we can continue to compute for $m = 2, 3, \ldots$ Here we only collect a few more results (Table 1). It is interesting to contrast these results with the case $\beta = 1$ (no discount) (Table 2).

Table 1

m	0	1	2	3	4	
$v_m(1)$	0	5.4	11.54	16.99	21.95	...
$v_m(2)$	15	24.05	29.90	35.34	40.25	...
$g_m(1)$	2	2	1	2	2	...
$g_m(2)$	2	1	1	1	1	...

Table 2

m	0	1	2	3	4	
$v_m(1)$	0	6	12.85	19.72	26.59	...
$v_m(2)$	15	24.5	31.55	38.46	45.34	...
$g_m(1)$	2	2	1	1	1	...
$g_m(2)$	2	1	1	1	1	...

□

(*d*) *Deterministic dynamical optimization.* If for every state i and every action a the next state j is determined by a deterministic function $j = s(i, a)$ we have $p_{ij}^a = 1$ for $j = s(i, a)$ and $p_{ij}^a = 0$ for $j \neq s(i, a)$ for every $i = 1, 2, \ldots, N$ and all $a \in A_i$. Thus the 'motion' of the system is purely *deterministic*. This important special case, too, falls into the category of Markovian decision problems. As an example we may consider the replacement problem, Example (*b*), if $p_i = 1$ for all $i = 0, 1, \ldots, N$; i.e. if the probability of a failure $1 - p_i$ is negligibly small. In this case $s(i, n) = i + 1$ and $s(i, k) = k + 1$, $k = 0, 1, \ldots, N - 1$.

For deterministic problems the backward recursion (10) of dynamical optimization takes the form

$$v_0(i) = \max_{a \in A_i} \{r_i^a\},$$

$$v_m(i) = \max_{a \in A_i} \{r_i^a + \beta v_{m-1}(s(i, a))\}, \quad i = 1, \ldots, N.$$

□

If we let the number of periods or steps m up to the end of the planning horizon tend to infinity, we obtain an *infinite-step* decision problem. The following theorem tells us about the convergence of the $v_m(i)$.

Theorem 3

For $0 \leqslant \beta < 1$ the boundary values $\lim_{m \to \infty} v_m(i) = v(i)$, $i = 1, 2, \ldots, N$, exist and are the uniquely determined solutions of the equations

$$v(i) = \max_{a \in A_i} \left\{ r_i^a + \beta \sum_{j=1}^{N} p_{ij}^a v(j) \right\}, \quad i = 1, \ldots, N. \tag{11}$$

Proof. We define a map T of the N-dimensional space into itself by

$$x_i = \max_{a \in A_i} \left\{ r_i^a + \beta \sum_{j=1}^{N} p_{ij}^a y_j \right\}, \quad i = 1, \ldots, N. \tag{12}$$

If v_m is the vector with components $v_m(i)$ we have, according to (10), $v_m = Tv_{m-1}$.

For the N-dimensional space we use the standard norm $\|x - y\| = \max_{i=1,\ldots,N} |x_i - y_i|$ and show first that $\|Tx - Ty\| \leqslant \beta \|x - y\|$. A map which has this property is called a *contraction*.

For two N-dimensional vectors x and y, let

$$(Tx)_i = r_i^{a'} + \beta \sum_{i=1}^{N} p_{ij}^{a'} x_j, \quad (Ty)_i = r_i^{a''} + \beta \sum_{j=1}^{N} p_{ij}^{a''} y_j,$$

where a' and a'' are elements of A_i which *maximize* (12) if applied to x and y. It follows that

$$(Tx)_i - (Ty)_i \leqslant r_i^{a'} + \beta \sum_{j=1}^{N} p_{ij}^{a'} x_j - r_i^{a'} - \beta \sum_{j=1}^{N} p_{ij}^{a'} y_j$$

$$\leqslant \beta \sum_{j=1}^{N} p_{ij}^{a'} \max_{i=1,\ldots,N} |x_i - y_i| = \beta \|x - y\|.$$

Similarly we have, with a' instead of a'', $(Tx)_i - (Ty)_i \geqslant -\beta \|x - y\|$. Thus we have $|(Tx)_i - (Ty)_i| \leqslant \beta \|x - y\|$ for $i = 1, \ldots, N$ and therefore $\|Tx - Ty\| \leqslant \beta \|x - y\|$.

The second part of the proof is concerned with a well-known *fixed point theorem for contraction operators.* By repeated application of the inequality $\|Tx - Ty\| \leqslant \beta \|x - y\|$ it follows that $\|T^m x - T^m y\| \leqslant \beta^m \|x - y\| \to 0$ as $m \to \infty$. If we now consider a sequence $z_m = Tz_{m-1}$, then $\|z_0 - z_m\| \leqslant \|z_0 - z_1\| + \|z_1 - z_2\| + \ldots + \|z_{m-1} - z_m\| \leqslant (1 + \beta + \ldots + \beta^{m-1}) \|z_0 - z_1\|$, i.e. $\|z_0 - z_m\|$ is bounded by $M = \|z_0 - z_1\|/(1 - \beta)$. Therefore we have $\|z_n - z_{n+m}\| = \|T^n z_0 - T^{n+m} z_0\| \leqslant \beta^n \|z_0 - T^m z_0\| \leqslant \beta^n M < \epsilon$ for all m if n is sufficiently large. Thus z_n is a *Cauchy sequence* and $\lim_{n \to \infty} z_n$ exists. In particular $\lim_{n \to \infty} v_n = v$ exists. Since $\|T^m x - T^m y\| \to 0$ as $m \to \infty$, all sequences $z_m = Tz_{m-1}$ must have the *same* limit v and if we take the limit on both sides of $z_m = Tz_{m-1}$, we obtain $v = Tv$. This relation is nothing but (11). A vector v which satisfies $v = Tv$ is called a *fixed point* of the map T. There cannot be two distinct fixed points v and w, since $\|v - w\| = \|Tv - Tw\| \leqslant \beta \|v - w\|$ and $\beta < 1$ together imply that $v = w$. Thus v is the unique fixed point of T. The theorem is proved. ∎

Infinite-step Markov decision problems are further examined in the next section.

5.2. Infinite-step Markov decision problems with discounting

In this section we shall consider infinite-step Markov decision problems as introduced at the end of the last section. Let us again summarize the elements of the problem: a system can be in states $i \in I = \{1, 2, \ldots, N\}$. If the system is in state i at time n we have to choose an action $a \in A_i$. All the A_i are finite sets. We are given an *initial distribution* p_i, $i \in I$, of the states and *transition probabilities* p_{ij}^a, $i,j \in I$, $a \in A_i$, for the transition from state i to state j for an action a. The sum of the p_{ij}^a over j is equal to 1 for every $i \in I$, $a \in A_i$. Moreover, pay-offs r_i^a are defined if the system is in state $i \in I$ and an action $a \in A_i$ has been chosen. The pay-offs at time n are discounted with a *discount factor* $0 \le \beta < 1$ for time 0 such that in imitation of Section 5.1 we have $r_{ni}^a = \beta^n r_i^a$.

As suggested by Section 5.1 for Markov decision problems we shall only consider *Markovian* policies $f = \{f_0, f_1, f_2, \ldots\}$ where the $f_n : I \to A = A_1 \cup \ldots \cup A_N$ are maps from the state space into the action set A with $f_n(i) \in A_i$. Let $r(f_n)$ be the vector of the pay-offs with components r_i^a, $a = f_n(i)$, $i \in I$, p the vector of the initial distributions with components p_i, $i \in I$, and let $P(f_n)$ be the matrix of the transition probabilities with elements p_{ij}^a, $a = f_n(i)$, $i, j \in I$. This matrix is *stochastic* for all f_n.

Every Markovian policy $g = \{g_0, g_1, \ldots\}$ determines an — in general *inhomogeneous* — Markov chain; see Section 5.1. Let $Q_n(g)$ be the matrix of the *n-step transition probabilities* $P[X_n = j | X_0 = i]$, $i, j \in I$ of the inhomo- geneous Markov chain $\{X_n, n = 0, 1, 2, \ldots\}$ which is determined by the policy g. From the Chapman–Kolmogorov equations (3.1.7), applied to $t = 0, m = 1, n - 1$, it follows that $Q_n(g) = Q_{n-1}(g)P(g_{n-1})$. Hence $Q_n(g) = P(g_0)P(g_1) \ldots P(g_{n-1})$, $n = 1, 2, \ldots$. Here we put $Q_0(g) = I$ (identity matrix). The expected value of the total pay-offs for the first $h + 1$ periods $n = 0, 1, 2, \ldots, h$, using policy g with initial state i, and discounted for time 0, is equal to $F_{hg}^h(i)$ in the notation of Section 5.1. Let F_g^h be the vector with components $F_{hg}^h(i)$. The expected value of the pay-offs at time n discounted for time 0 is $\beta^n Q_n(g)r(g_n)$, and therefore

$$F_g^h = \sum_{n=0}^{h} \beta^n Q_n(g)r(g_n), \quad h = 0, 1, 2, \ldots . \tag{1}$$

The sums (1) *converge* as $h \to \infty$ since they are majorized by convergent geometric series for $0 \le \beta < 1$. Let $F_g = \lim_{h \to \infty} F_g^h$ be the vector of the pay-offs over *infinitely many* periods under policy g, discounted for time 0,

$$F_g = \sum_{n=0}^{\infty} \beta^n Q_n(g)r(g_n). \tag{2}$$

Let the components of F_g be denoted by $F_g(i)$. A policy g^* is called β-optimal if $F_{g*}(i) \geqslant F_g(i)$ for all $i = 1, 2, \ldots, N$ and all policies g. The *infinite-step Markov decision problem with discounting* consists of determining β-optimal policies. If G_g is the expected pay-off under policy g with initial distribution p, discounted for time 0,

$$G_g = \sum_{i=1}^{N} p_i F_g(i), \tag{3}$$

and if g^* is a β-optimal policy, we also have $G_{g*} \geqslant G_g$ for every policy g.

Examples

Examples (5.1.a) and (5.1.b) of maintenance and replacement problems, as well as the general deterministic dynamical optimization of Example (5.1.d) can also be considered with infinite planning horizon, and then fall into the category of infinite-step Markov decision problems defined above. Let us consider a further example.

(a) *Stock-keeping.* Suppose the stock of an item is in the range of $i = 0, 1, \ldots, N$ pieces. The demands X_n per period are mutually independent and distributed with $P[X_n = j] = q_j, j = 1, 2, \ldots$. At any time the stock can be increased to the maximum N. The permissible actions at a level i are thus $a = 0, 1, \ldots, N - i$, where a represents the quantity ordered for re-stocking. If the demand is higher than the existing stock after re-stocking, then the excess demand is lost owing to lack of supply.

The transition probabilities for this system are defined as follows:

$$p_{ij}^a = 0, \qquad \text{if } i + a < j,$$

$$p_{ij}^a = q_{i+a-j}, \qquad \text{if } 0 < j \leqslant i + a,$$

$$p_{ij}^a = \sum_{k \geqslant i+a} q_k, \qquad \text{if } j = 0.$$

The following costs have to be considered: (a) ordering or production costs $c_1(a)$ depending on the quantities ordered; (b) stock-keeping costs, which are basically interest costs on the capital tied up in the stock. In the simplest case one may assume these to be proportional to the level i of stock, $c_2 i$. (c) Costs for k lost sales $c_3(k)$. If the level of stock is i and the quantity to be ordered is a, then the expected value of the cost of lost sales (deficiency cost) per unit time is

$$\bar{c}_3(i, a) = \sum_{k > i+a} c_3(k - i - a) q_k.$$

We can thus define $r_i^a = c_1(a) + c_2 i + \bar{c}_3(i, a)$. An order policy has to be determined which *minimizes* the discounted costs over infinitely many periods.

The minimizing problem can as usual be translated into a maximizing problem by considering the pay-offs $-r_i^a$. □

Let F be the set of all maps $f: I \to A$ with $f(i) \in A_i$. If $g = \{g_0, g_1, g_2, \ldots\}$, $g_i \in F$, is a policy for $i = 0, 1, 2, \ldots$ and if $f_0, f_1, \ldots, f_m \in F$, then let $(f_0, f_1, \ldots, f_m, g)$ be the policy $\{h_n\}$ with $h_n = f_n, n = 0, 1, \ldots, m$ and $h_n = g_{n-m-1}, n = m + 1, m + 2, \ldots$. If $f \in F$, then f^∞ is the policy $\{h_n\}$ with $h_n = f, n = 0, 1, 2, \ldots$; f^∞ is called a *stationary* policy. Finally, let Dg denote the policy $\{h_n\}$ with $h_n = g_{n+1}, n = 0, 1, 2, \ldots$, where g is an arbitrary policy. For stationary policies f^∞ we have $Df^\infty = f^\infty$. A stationary policy determines a *homogeneous* Markov chain.

It follows from (2) that

$$F_g = r(g_0) + \beta P(g_0) \sum_{n=0}^{\infty} \beta^n Q_n(Dg) r(g_{n+1})$$

$$= r(g_0) + \beta P(g_0) F_{Dg}. \tag{4}$$

For all $f \in F$, let $T(f)$ be a map which transforms the N-dimensional vectors v into $T(f)v = r(f) + \beta P(f)v$. We write $v_1 \geqslant v_2$ if all components of v_1 are greater than or equal to the corresponding ones of v_2 and $v_1 > v_2$, if $v_1 \geqslant v_2$ and $v_1 \neq v_2$. If g and f are two policies, we write $g \geqslant f$ ($g > f$), if $F_g \geqslant F_f$ ($F_g > F_f$).

Theorem 1

(a) If g is a policy for which we have $g \geqslant (f, g)$ for all $f \in F$, then g is β-*optimal*. (b) Conversely, if $(f, g) > g$ for an $f \in F$, then we also have $f^\infty > g$.

Proof. (a) $D(f, g) = g$ and thus, by (4), and by assumption,

$$F_{(f,g)} = T(f)F_g \leqslant F_g. \tag{5}$$

For every arbitrary policy h we have thus $T(h_n)F_g \leqslant F_g$. Let us now assign to the policy h the policy $h^n = (h_0, h_1, h_2, \ldots, h_n, g)$. We then have $g = D^{n+1}h^n$ and by repeated application of (4) it follows that

$$F_{h^n} = T(h_0)T(h_1)\ldots T(h_n)F_g = \sum_{m=0}^{n} \beta^m Q_m(h)r(h_m) + \beta^{n+1}Q_{n+1}(h^n)F_g.$$

Because of the monotonicity of $T(f)$ it follows from $T(h_n)F_g \leqslant F_g$ that

$$F_{h^n} = T(h_0)T(h_1)\ldots T(h_n)F_g \leqslant F_g, \quad n = 0, 1, 2, \ldots,$$

and therefore

$$F_g \geqslant \lim_{n \to \infty} F_{h^n} = \lim_{n \to \infty} \left\{ \sum_{m=0}^{n} \beta^m Q_m(h)r(h_m) + \beta^{n+1}Q_{n+1}(h^n)F_g \right\} = F_h.$$

g is thus β-optimal.

(*b*) By assumption and by (5), $T(f)F_g > F_g$, and therefore because of the monotonicity of $T(f)$, $T^n(f)F_g \geqslant T^{n-1}(f)F_g$ and hence $T^n(f)F_g > F_g$. From this it follows (with $P^0(f) = I$) that

$$\lim_{n \to \infty} T^n(f)F_g = \lim_{n \to \infty} \left\{ \sum_{m=0}^{n} \beta^m P^m(f)r(f) + \beta^{n+1}P^{n+1}(f)F_g \right\} = F_{f^\infty} > F_g,$$

and we have proved $f^\infty > g$. ∎

On the basis of Theorem 1 we can in the following obtain a major result of the theory of infinite-step Markov decision processes with discounting: there must exist a β-optimal policy which is *stationary*. Intuitively this is very plausible. After the first period exactly the same infinite-step decision problem appears. Why should one therefore apply a different decision rule at the second and all subsequent periods from the first one? The proof that there exists a stationary *β-optimal* policy is given constructively and in such a way that we immediately obtain a computational method with which we can determine a β-optimal stationary policy.

Theorem 2
    Let $f \in F$ and let $G(i,f)$ for $i = 1, 2, \ldots, N$ be the set of all $a \in A_i$, for which we have

$$r_i^a + \beta \sum_{j=1}^{N} p_{ij}^a F_{f^\infty}(j) > F_{f^\infty}(i), \tag{6}$$

where $F_{f^\infty}(i) = 1, 2, \ldots, N$ are the components of F_{f^∞}. Then f^∞ is β-optimal if and only if $G(i,f)$ is empty for all $i = 1, 2, \ldots, N$.

Proof. If $g \in F$ we have, provided that all $G(i,f)$ are empty,

$$F_{(g,f^\infty)} = r(g) + \beta P(g)F_{f^\infty} \leqslant F_{f^\infty}.$$

Therefore $(g, f^\infty) \leqslant f^\infty$ for all $g \in F$ and f^∞ is *β-optimal* according to Theorem 1(*a*).
    If on the other hand not all $G(i,f)$ are empty we can determine a $g \in F$ as follows: $g(i) = a$ with $a \in G(i,f)$ if $G(i,f)$ is not empty, $g(i) = f(i)$ if $G(i,f)$ is empty. For the policy (g, f^∞) we have, according to (6),

$$F_{(g,f^\infty)} = r(g) + \beta P(g)F_{f^\infty} > F_{f^\infty}$$

and $(g, f^\infty) > f^\infty$. Thus f^∞ is not β-optimal if not all $G(i,f)$ are empty. ∎

It follows from (4) and $f^\infty = Df^\infty$ that Theorem 2 is equivalent to the following corollary.

Corollary
    f^∞ is β-optimal if and only if we have

$$F_{f^\infty}(i) = \max_{a \in A_i} \left\{ r_i^a + \sum_{j=1}^{N} p_{ij}^a F_{f^\infty}(j) \right\}, \quad i = 1, \ldots, N. \tag{7}$$

The equations (7) are called *optimality equations* or *functional equations* of the dynamical optimization.

The equations (7) are precisely of the form (5.1.11) which will enable us to connect the methods of Section 5.1 with the infinite-step decision problem. First, however, there still remains the question of the existence of an optimal stationary policy f^∞ which will have to satisfy (7). If there exists such a policy, then, according to (4) and because of $f^\infty = Df^\infty$, F_{f^∞} has to be a solution of the equations

$$v = r(f) + \beta P(f)v \tag{8}$$

and $a = f(i)$ has to maximize (7).

To continue from this remark we are able to construct a method with which we can determine an optimal stationary policy. For this we need the following theorem:

Theorem 3
The system of equations (8) has the *unique* solution $v = F_{f^\infty}$.

Proof. For $0 \le \beta < 1$ the series $I + \beta P(f) + \beta^2 P^2(f) + \ldots$ converges and is equal to $(I - \beta P(f))^{-1}$ as we can verify by multiplying the sequence with $(I - \beta P(f))$. (8) has thus the unique solution $v = (I - \beta P(f))^{-1} r(f)$, and since F_{f^∞} is the solution of (8), $v = F_{f^\infty}$ must hold. ∎

We can thus try to find a policy f^∞ satisfying (7) in the following manner: one fixes an $f \in F$ and goes through the following steps:
1. *Value determination*: Solve (8) in order to determine $v = F_{f^\infty}$.
2. *Policy improvement*: Determine the sets $G(i, f)$ according to Theorem 2.

If all $G(i, f)$ are empty, then, according to Theorem 2, f^∞ is β-optimal. If not, we choose, for all i with non-empty $G(i, f)$, an $a \in G(i, f)$ and put $g(i) = a$. For all i for which the $G(i, f)$ are empty, we put $g(i) = f(i)$. Then we repeat steps 1 and 2 with $g = (g(1), \ldots, g(N)) \in F$.

This method is called *policy iteration*. The following theorem shows that we can always find a β-optimal stationary policy by the policy iteration method. If we want to minimize instead of maximize, all we have to do is to reverse the direction of the inequality in (6) and replace max by min in (7).

Theorem 4
(a) If g^∞ is the new stationary policy which we obtain by the policy iteration method from f^∞, then $g^\infty > f^\infty$.

(*b*) The policy iteration method will terminate after finitely many steps with a stationary β-optimal policy.

(*c*) There always exists a stationary β-optimal policy.

Proof. (*a*) According to (4) and by assumption we have

$$F_{(g,f^\infty)} = r(g) + \beta P(g)F_{f^\infty} > F_{f^\infty}.$$

Therefore $(g,f) > f^\infty$ and thus, according to Theorem 1(*b*), $g^\infty > f^\infty$. The second step in the policy iteration method is thus rightly called a policy improvement.

(*b*) According to (*a*) each new policy in the policy iteration method is better than the previous one. Since there are only finitely many stationary policies, this method has to terminate after finitely many steps. According to Theorem 2 the stationary policy with which the method will terminate is β-optimal.

(*c*) follows immediately from (*b*). ∎

Examples

(*b*) *A computational example.* Let us take up Example (5.1.*c*) once more and solve the associated infinite-step decision problem with the help of policy iteration. As in Example (5.1.*c*) we choose the discount factor $\beta = 0.9$.

As an initial policy for the iteration let us choose the policy $f(1) = f(2) = 1$. The equations (8) for the determination of the value are

$$v_1 = 5 + 0.81\,v_1 + 0.09\,v_2,$$

$$v_2 = 20 + 0.63\,v_1 + 0.27\,v_2.$$

These equations have the solution $v_1 = 66.463\,27$, $v_2 = 84.7558$. For a policy improvement the left hand sides of (6) have to be computed for $a = 2$ and the states $i = 1, 2$

$$0 + 0.54 \times 66.463\,27 + 0.36 \times 84.7558 = 66.40 < 66.463\,27,$$

$$15 + 0.18 \times 66.463\,27 + 0.72 \times 84.7558 = 87.99 > 84.7558.$$

Therefore $G(1, 1) = \{2\}$ and $G(1, 2)$ is empty (we are dealing here with a minimization problem). The new policy $f(1) = 2, f(2) = 1$ has thus to be better than the old one.

The values of the new policy arise from the equations

$$v_1 = 0 + 0.54\,v_1 + 0.36\,v_2,$$

$$v_2 = 20 + 0.63\,v_1 + 0.27\,v_2.$$

This system has the solution $v_1 = 66.06$, $v_2 = 84.40$.

For a policy improvement we have to compute the left hand sides of (6) for state 1 and action 1 as well as for state 2 and action 2

$$5 + 0.81 \times 66.06 + 0.09 \times 84.40 = 66.10 > 66.06,$$

$$15 + 0.18 \times 66.06 + 0.72 \times 84.40 = 87.66 > 84.40.$$

According to this it is impossible to improve the policy any further. The stationary policy with $f(1) = 2, f(2) = 1$ is *optimal.*   □

(*c*) *Replacement problems.* For the discounted infinite-step variant of the replacement problem, Example (5.1.*b*), the equations (8) for the determination of value read as follows:

$$v_i = B_i + \beta(p_i v_{i+1} + (1 - p_i)v_N) \qquad \text{if } f(i) = n,$$

$$v_i = C_k - T_i + B_k + \beta(p_k v_{k+1} + (1 - p_k)v_N) \qquad \text{if } f(i) = k.$$

For a policy improvement we can choose the action a which minimizes

$$B_i + \beta(p_i v_i + (1 - p_i)v_N)$$

and

$$\min_{k=0,1,\ldots,N-1} \{C_k - T_i + B_k + \beta(p_k v_{k+1} + (1 - p_k)v_N)\}.$$

We obtain the minimum of the last expression for all i for the *same* action $k = m$. Therefore in policy iteration we only have to consider policies of the form $g(i) = n$ or m. *Either no replacement takes place at all or if a replacement does take place, it will be independent of age i for a machine of age m!* In particular there exists a β-optimal policy of this form. This remark allows us to simplify the policy improvement since the above minimization over $k = 1, \ldots, N - a$ has only to be done once for every i. For some concrete problems the special structure allows such a simplification.   □

(*d*) *Deterministic infinite-step decision problem.* The policy iteration method can also be applied to the infinite-step version of the deterministic dynamical optimization, Example (5.1.*d*). The equations (8) for the determination of value take on the form

$$v(i) = r_i^{f(i)} + \beta v(s(i, f(i))), \quad i = 1, \ldots, N,$$

and the functional equation (7) reads

$$F_{f^\infty}(i) = \max_{a \in A_i} \{r_i^a + \beta F_{f^\infty}(s(i, a))\}, \quad i = 1, \ldots, N.$$

For the policy improvement in the policy iteration method we consider the sets $G(i, f)$ of all $a \in A_i$ for which we have

$$r_i^a + \beta v(s(i, a)) > v(i), \quad i = 1, \ldots, N.$$ □

Theorem 5.1.3 permits the introduction of a second computational method besides the policy iteration with which we can determine β-optimal policies. If we apply the backward recursion of dynamical optimization for $m = 1, 2, \ldots$ we obtain the maximum discounted total pay-offs $v_m(i)$ over

m periods by (5.1.10) which, according to Theorem 5.1.3, converge for $m \to \infty$. The limits satisfy the functional equations (7) and are therefore equal to the discounted pay-offs of the β-*optimal* policies. For $m = 0, 1, 2, \ldots$, let $f_m(i) \in A_i$ be the actions which maximize (5.1.10) and let $f_m = (f_m(1), \ldots, f_m(N))$. The following Theorem 5 will show that f_m^∞ is β-optimal for all sufficiently large m. It is therefore sufficient to carry out the backward recursion of the dynamical optimization for a sufficiently large m in order to obtain a β-optimal policy. This method is called *value iteration*.

Theorem 5
Let $f_m(i) \in A_i$ be an action which *maximizes*

$$v_m(i) = \max_{a \in A_i} \left\{ r_i^a + \beta \sum_{j=1}^N p_{ij}^a v_{m-1}(j) \right\}, \quad v_{-1}(i) = 0,$$

$$i = 1, \ldots, N; \quad m = 0, 1, 2, \ldots, \tag{9}$$

and let $f_m = (f_m(1), \ldots, f_m(N))$. Then f_m^∞ is β-optimal for all sufficiently large m.

Proof. Let $B_i \subset A_i$ be the set of actions which maximizes the functional equations

$$v(i) = \max_{a \in A_i} \left\{ r_i^a + \beta \sum_{j=1}^N p_{ij}^a v(j) \right\}.$$

These functional equations have, according to Theorem 5.1.3, uniquely determined solutions $v(j)$. For a policy f^∞ with $f = (f(1), \ldots, f(N)), f(i) \in B_i$, we have (7) and such a policy is β-optimal, according to the corollary of Theorem 2. It is therefore sufficient to show that $f_m(i) \in B_i$ for sufficiently large m.

There exists an $\epsilon > 0$ such that

$$r_i^a + \beta \sum_{j=1}^N p_{ij}^a v(j) > r_i^{a'} + \beta \sum_{j=1}^N p_{ij}^{a'} v(j) + 2\epsilon \tag{10}$$

for all $a \in B_i$ and for $a' \in A_i - B_i$. Furthermore, by Theorem 5.1.3, there exists an $n(\epsilon)$ such that we have for all $m > n(\epsilon)$ and $j = 1, \ldots, N$, $|v_m(j) - v(j)| < \epsilon$. For all $a \in A_i$ and for all $m - 1 > n(\epsilon)$,

$$r_i^a + \beta \sum_{j=1}^N p_{ij}^a v(j) - \epsilon < r_i^a + \beta \sum_{j=1}^N p_{ij}^a v_{m-1}(j) < r_i^a + \beta \sum_{j=1}^N p_{ij}^a (j) v(j) + \epsilon.$$

$$\tag{11}$$

If $a'' = f_m(i)$ is an action which maximizes (9), it follows from (11) that

$$r_i^a + \beta \sum_{j=1}^{N} p_{ij}^a v(j) - \epsilon < r_i^a + \beta \sum_{j=1}^{N} p_{ij}^a v_{m-1}(j)$$

$$\leqslant r_i^{a''} + \beta \sum_{j=1}^{N} p_{ij}^{a''} v_{m-1}(j) < r_i^{a''} + \beta \sum_{j=1}^{N} p_{ij}^{a''} v(j) + \epsilon.$$

This holds in particular for $a \in B_i$ and, by (10), we have to have $a'' = f_m(i) \in B_i$ and f_m^∞ is β-optimal.   ∎

Compared with policy iteration, value iteration has the advantage that the — for large N — laborious solution of the equations (8) for the determination of value can be avoided. In the method of value iteration we lack, however, a criterion which allows us to determine when m is sufficiently large. We can, of course, stop the value iteration at an arbitrary m and choose f_m^∞ as the initial policy for policy iteration and hope that this process will terminate with the first step and that f_m^∞ is thus already β-optimal. The methods of policy iteration and value iteration can therefore be combined.

Example

(e) *Computational example.* When comparing Examples (b) and (5.1.c) we can see that value iteration has already reached the β-optimal policy for the infinite-step decision problem for $m = 3$ (see the table in Example 5.1.c)). However, $v_m(1)$ and $v_m(2)$ converge slowly to v_1 and v_2. According to Example (b) $v_1 = 66.06$, $v_2 = 84.40$ whereas $v_4(1)$ and $v_4(2)$ are, according to Example (5.1.c), equal to 21.95 and 40.25. For smaller discount factors than $\beta = 0.9$ the convergence would be faster. An estimate of the speed of convergence is contained in the proof of Theorem 5.1.3.   □

5.3. Infinite-step Markov decision problems without discounting

In this section we consider the infinite-step Markov decision problem with $\beta = 1$. The case $\beta = 1$ (no discounting) is much more complicated than the case with discounting in the previous section since, for instance, Theorem 5.1.3 does not hold for $\beta = 1$. Therefore we shall refrain from dealing with the case $\beta = 1$ in full generality. In particular we shall limit our considerations *right from the start* to stationary policies f^∞, i.e. we shall look for the optimal policy only within the class of the stationary policies. Further restrictive assumptions are introduced below. We shall continue with the same notation as in the previous section.

Let us first define what we understand by an optimal policy in the case $\beta = 1$. The total pay-off over infinitely many periods will only be finite as an exception; (5.2.1) does not in general converge for $\beta = 1$. Therefore the notion of optimality of the last section cannot be sustained. A stationary policy f^∞ determines a *discrete homogeneous Markov chain* with transition matrix $P(f) = (p_{ij}^a), a = f(i)$, such that the n-step transition matrices are

$Q_n(f^\infty) = P^n(f), Q_0(f^\infty) = I$. In order to simplify the notation we shall in future just write f for the stationary policy instead of f^∞. Let F_f^h be the vector of the total pay-offs $F_f^h(i)$ in the initial state i for the policy f during the first $h + 1$ periods $n = 0, 1, 2, \ldots, h$. Let $r(f)$ be, as in Section 5.2, the vector of the pay-offs $r_i^f(i)$ in state i for policy f. We then have (see (5.2.1))

$$F_f^h = \sum_{n=0}^{h} P^n(f)r(f), \quad h = 0, 1, 2, \ldots . \tag{1}$$

In contrast to the case $\beta < 1$ the *class* structure of the Markov chain defined by $P(f)$ plays a fundamental role in the case $\beta = 1$. We make the following restrictive assumption: for every stationary policy f there exists a *unique recurrent and aperiodic class* and, possibly, *several additional transient states*. All transient states must, of course, lead to the states of the recurrent class. Whether a particular state is recurrent or transient generally depends on the policy to be considered.

Let us for the moment denote the elements of $P^n(f)$ by $p_{ij}^{(n)}(f)$. If j is *recurrent* we have, according to Theorem 3.2.4, $p_{ij}^n(f) \to p_j(f) > 0$ for $n \to \infty$. The $p_j(f)$ form, according to Theorem 3.2.9, is the unique stationary distribution for the class of recurrent states j. For *transient* states j we have, according to Theorem 3.2.4, $p_{ij}^{(n)}(f) \to p_j(f) = 0$ for $n \to \infty$. If $P^\infty(f)$ is the matrix with the elements $p_{ij}(f) = p_j(f), i, j = 1, 2, \ldots, N$, we can thus write $P^n(f) \to P^\infty(f)$ for $n \to \infty$. We obtain

$$P^\infty(f)P(f) = P(f)P^\infty(f) = P^\infty(f)P^\infty(f) = P^\infty(f). \tag{2}$$

The last two equalities hold because all rows of $P^\infty(f)$ are identical and $P^\infty(f), P(f)$ are stochastic. The first equality follows from (3.2.10). Since the arithmetic mean of a converging sequence converges to the same limit as the sequence itself it follows from (1) that

$$\lim_{h \to \infty} \frac{1}{h+1} F_f^h = P^\infty(f)r(f) = x(f), \tag{3}$$

where $x(f)$ has N identical components $x(f)$.

$x(f)$ is the expected *average* pay-off per period or, as (3) shows, also *the expected value of the pay-off under the stationary distribution*. $x(f)$ is called the *profit rate*. It certainly makes sense to optimize the profit rate $x(f)$ so that f is optimal if $x(f) \geqslant x(g)$ for all $g \in F$. Note that $x(f)$ is determined only by the pay-offs $r_i^{f(i)}$ in states i which are recurrent under the policy f.

We can make more precise statements about the asymptotic behaviour of the pay-offs.

Theorem 1
The limits

$$\lim_{h \to \infty} (F_f^h - (h+1)x(f)) = y(f) \tag{4}$$

exist, are finite and satisfy the equations

$$x(f) + y(f) = r(f) + P(f)y(f), \tag{5}$$

$$P^\infty(f)y(f) = 0. \tag{6}$$

To prove this theorem we need the following lemma from matrix theory.

Lemma

If A is an $N \times N$ matrix and $A^n \to 0$ for $n \to \infty$ then $I - A$ is non-singular and

$$(I - A)^{-1} = \sum_{n=0}^{\infty} A^n. \tag{7}$$

Proof of the lemma. For $0 < \epsilon < 1$ there exists an $n(\epsilon)$ such that for all $n > n(\epsilon)$, $|a_{ij}^n| < \epsilon/N, i,j = 1, \ldots, N$, if $a_{ij}^{(n)}$ is an element of A^n. Hence

$$|a_{ij}^{(2n)}| \leqslant \sum_{k=1}^{N} |a_{ik}^{(n)}| \, |a_{kj}^{(n)}| \leqslant \frac{\epsilon^2}{N}$$

and thus $|a_{ij}^{(sn)}| \leqslant \epsilon^s/N, s = 1, 2, 3, \ldots$. If an $n > n(\epsilon)$ is fixed, it follows for $m = 1, \ldots, n-1$ that

$$|a_{ij}^{(sn+m)}| \leqslant \sum_{k=1}^{N} |a_{ik}^{(sn)}| \, |a_{kj}^{(m)}| \leqslant \frac{M\epsilon^s}{N},$$

where

$$M = \max_{\substack{m=1,\ldots,n-1 \\ j=1,\ldots,N}} \left(\sum_{k=1}^{N} |a_{kj}^{(m)}| \right).$$

Hence the series in (7) converges absolutely in the elements $a_{ij}^{(n)}$ and multiplication of the series by $(I - A)$ shows that it is equal to $(I - A)^{-1}$. ∎

Proof of Theorem 1. According to (1) and (3) we have

$$F_f^h - (h+1)x(f) = \sum_{n=0}^{h} (P^n(f) - P^\infty(f))r(f). \tag{8}$$

It follows from (2) that $P^n(f)P^\infty(f) = P^\infty(f)$ and that $P^2(f) - P^\infty(f) = (P(f) - P^\infty(f))^2$ and hence, by induction on n, that $P^n(f) - P^\infty(f) = (P(f) - P^\infty(f))^n$ for $n = 1, 2, \ldots$ (but not for $n = 0$). Since $P^n(f) \to P^\infty(f)$ as $n \to \infty$ it follows from the lemma that the limits (4) exist and

$$y(f) = \left[\sum_{n=0}^{\infty} (P(f) - P^\infty(f))^n - P^\infty(f) \right] r(f)$$

$$= (I - P(f) + P^\infty(f))^{-1} r(f) - P^\infty(f)r(f). \tag{9}$$

From the first part of (9) follows, after multiplication by $P^{\infty}(f)$, relation (6) and from the second part and from (3) we obtain (5). ∎

Equation (4) states that $F_f^h(i)$ approaches asymptotically, for all $i = 1, \ldots, N$ and for large h, a straight line with inclination $x(f)$. For the states $i = 1, 2, \ldots, N$ the components $y(i, f)$ of $y(f)$ denote the *difference in the second coordinate* between this asymptotic straight line and the straight line $y = (h + 1)x(f)$; the $y(i, f)$ are called *bias values*.

Equation (5) has the form of a system of linear equations

$$x + v_i = r_i + \sum_{j=1}^{N} p_{ij} v_j, \quad i = 1, \ldots, N, \tag{10}$$

with N equations for $N + 1$ unknowns x and $v_i, i = 1, \ldots, N$. The following theorem describes the solutions of these equations.

Theorem 2

Suppose that the matrix of the transition probabilities p_{ij} in (10) defines a Markov chain with a unique recurrent class and, possibly, several additional transient states (in accordance with the standing assumptions in this section). Then x is *uniquely* defined by (10) while the $v_i, i = 1, 2, \ldots, N$, are uniquely defined by (10) *up to an additive constant.*

Proof. According to the assumptions the transition matrix p_{ij} has a unique stationary distribution $p_i, i = 1, 2, \ldots, N$. If we multiply (10) by p_i and sum over i we obtain, by observing (3.2.10),

$$x = \sum_{i=1}^{N} p_i r_i \tag{11}$$

and x is determined uniquely as the *profit rate.*

If $x, v_i, i = 1, 2, \ldots, N$ is a solution of (10), then $x, v_i + a, i = 1, 2, \ldots, N$, where a is an arbitrary constant, is obviously also a solution of (10). It only remains to be shown that any two solutions v_i' and v_i'' of (10) can only differ by a constant, $v_i' - v_i'' = a$ for all $i = 1, 2, \ldots, N$. The differences $w_i = v_i' - v_i''$ satisfy the equations

$$w_i = \sum_{j=1}^{N} p_{ij} w_j. \tag{12}$$

If we multiply (12) by p_{ki} and sum over i we realize that (12) also has to hold for $p_{ij}^{(2)}$ instead of p_{ij} and therefore more generally, for

$$w_i = \sum_{j=1}^{N} p_{ij}^{(n)} w_j, \quad n = 1, 2, \ldots . \tag{13}$$

If we let $n \to \infty$, it follows that (see Theorems 3.2.4 and 3.2.9)

$$w_i = \sum_{j=1}^{N} p_j w_j. \tag{14}$$

The right hand side is, however, independent of i, and all w_i have to have the same value.   ∎

The bias values $y(f)$ are thus not uniquely determined by (5) and we need the additional condition (6). The differences $v_i - v_j$, however, are unique according to Theorem 2, and indicate the limits of the differences of the total pay-offs from state i and from state j as $h \to \infty$. Therefore the solutions v_i of (10) are also called *relative values*. We can often refrain from computing the bias values and only determine the relative values of a policy from (5) instead, by setting the relative value of an arbitrary state, for instance v_N, equal to zero. In this case the stationary distribution appearing in (6) need not be computed.

In the case without discounting there also exists an analogue to Theorem 5.2.2.

Theorem 3
Let $f \in F$ and let $G(i,f)$ for $i = 1, 2, \ldots, N$ be the set of all $a \in A_i$ for which

$$r_i^a + \sum_{j=1}^{N} p_{ij}^a v(j,f) > x(f) + v(i,f), \tag{15}$$

where the $v(i,f)$ are the relative values of the policy f. If $G(i,f)$ is empty for all $i = 1, 2, \ldots, N$, then f is optimal

Proof. First note that the sets $G(i,f)$ are independent of the choice of the relative values in (15). Now let g be a second policy with a profit rate $x(g)$ and relative values $v(i,g)$ and let

$$d_i = r_i^{f(i)} + \sum_{j=1}^{N} p_{ij}^{f(i)} v(j,f) - r^{g(i)} - \sum_{j=1}^{N} p_{ij}^{g(i)} v(j,f). \tag{16}$$

The first expression in (16) on the right hand side is, by (10), equal to $x(f) + v(i,f)$. Therefore $d_i < 0$ if $g(i) \in G(i,f)$, and $d_i \geqslant 0$ otherwise. If we subtract the equations

$$x(f) + v(i,f) = r_i^{f(i)} + \sum_{j=1}^{N} p_{ij}^{f(i)} v(j,f),$$

$$x(g) + v(i,g) = r_i^{g(i)} + \sum_{j=1}^{N} p_{ij}^{g(i)} v(j,g),$$

if we write $\Delta x = x(f) - x(g)$, $\Delta v_i = v(i,f) - v(i,g)$ and if we substitute for $r_i^{f(i)} - r_i^{g(i)}$, from (16) we obtain

$$\Delta x + \Delta v_i = d_i + \sum_{j=1}^{N} p_{ij}^{g(i)} \Delta v_j. \tag{17}$$

(17) has the form of (10). If p_i^g is the stationary distribution for $(p_{ij}^{g(i)})$, we have, according to (11),

$$\Delta x = \sum_{i=1}^{N} p_i^g d_i. \tag{18}$$

If all $G(i, f)$ are empty, then $d_i \geqslant 0$ for $i = 1, \ldots, N$ and therefore $\Delta x \geqslant 0$. Hence f is optimal. ∎

In contrast to Theorem 5.2.2, Theorem 3 gives a sufficient but not a necessary condition: if g is an optimal policy for which a state i is transient, i.e. $p_i^g = 0$, it is quite possible for a policy f to have $d_i < 0$ and $d_j = 0$ for $j \neq i$. According to (18), $\Delta x = 0$ and f is also optimal, although $G(i, f)$ is not empty. We shall return to this later.

As in Section 5.2 we have

Corollary
If for an $f \in F$ with profit rate $x(f)$ and relative values $v(i, f)$ we have

$$x(f) + v(i, f) = \max_{a \in A_i} \left\{ r_i^a + \sum_{j=1}^{N} p_{ij}^a v(j, f) \right\}, \tag{19}$$

then f is optimal. Equations (19) are the *optimality* or *functional equations* for the case without discounting, considered here.

As in the case with discounting we can set up a completely analogous *policy iteration method*: for any $f \in F$ the following computational steps are used:

1. *Determination of value.* Solve the equations

$$x + v_i = r_i^{f(i)} + \sum_{j=1}^{N} p_{ij}^{f(i)} v_j, \quad i = 1, \ldots, N, \tag{20}$$

   in order to determine the profit rate $x(f)$ and the relative values $v_i = v(i, f)$ of the policy f, where an arbitrary relative value, for instance v_N, is put equal to zero.
2. *Policy improvement.* Determine the sets $G(i, f)$ according to Theorem 3. If all $G(i, f)$ are empty, then, according to Theorem 3, f is already optimal. If not, we choose for all i for which $G(i, f)$ is not empty an $a \in G(i, f)$ and put $g(i) = a$. For all i for which $G(i, f)$ is empty put $g(i) = f(i)$. We then repeat steps 1 and 2 with $g = (g(1), \ldots, g(N)) \in F$.

The properties of the policy iteration method are collected in the following theorem:

Theorem 4

Let g be the new policy obtained by the policy iteration method from f. Let, furthermore, R^g and T^g be the sets of *recurrent* and *transient* states under the policy g. We have:

(a) If $G(i, f)$ is *not empty* for at least one recurrent state $i \in R^g$, then $x(g) > x(f)$.

(b) If $G(i, f)$ is *empty* for all recurrent states $i \in R^g$, then $x(g) = x(f)$. If we put $v(j, f) = v(j, g)$ for some $j \in R^g$, then we have $v(i, f) = v(i, g)$ for *all* $i \in R^g$. If, however, $G(i, f)$ is *not empty* for some $i \in T^g$, we have $v(i, g) \geqslant v(i, f)$ for all $i \in T^g$ and $v(i, g) > v(i, f)$ for at least one $i \in T^g$.

Proof. (a) Let $d_i, i = 1, \ldots, N$, be defined by (16) where f is the old and g the new policy in the policy iteration. As in the proof of Theorem 3 it follows that

$$\Delta x = x(f) - x(g) = \sum_{i=1}^{N} p_i^g d_i, \tag{21}$$

where p_i^g is the stationary distribution of the Markov chain for the policy g. We have $p_i^g = 0$ for $i \in T^g$ and $p_i^g > 0$ for $i \in R^g$. If for an $i \in R^g$ the set $G(i, f)$ is non-empty, then $d_i < 0$ and, since $p_i^g > 0$, it follows that $\Delta x < 0$ or $x(g) > x(f)$.

(b) If the $G(i, f)$ are empty for all $i \in R^g$, then $d_i = 0$ for all $i \in R^g$ and since $p_i^g = 0$ for all $i \in T^g$, it follows from (21) that $\Delta x = 0$ or $x(g) = x(f)$. If we introduce the column vectors Δv and d with the components $v(i, f) - v(i, g)$ and d_i, respectively, equations (17) take on the form

$$(I - P(g))\Delta v = d. \tag{22}$$

Let the states $i = 1, \ldots N$ be numbered in such a way that $T^g = \{1, \ldots, m\}$ and $R^g = \{m + 1, \ldots, N\}$. We decompose the matrix $I - P(g)$ in (22) corresponding to the sets T^g and R^g into

$$I - P(g) = \begin{bmatrix} A & B \\ 0 & C \end{bmatrix}. \tag{23}$$

Here $A = (\delta_{ij} - p_{ij}^{g(i)}), i, j \in T^g; B = (-p_{ij}^{g(i)}), i \in T^g, j \in R^g; C = (\delta_{ij} - p_{ij}^{g(i)})$, $i, j \in R^g$, $(\delta_{ij} = 0$ for $i \neq j, \delta_{ij} = 1$ for $i = j)$. The zero matrix appears on the bottom left hand side since a recurrent state cannot lead to a transient state (Theorem 3.2.6), and hence $p_{ij}^{g(i)} = 0$ for $i \in R^g, j \in T^g$.

Equation (22) is a special case of (17) and (17) is a system of equations of the form (10). Exactly as in (10) we can put in (22) one component of Δv equal to zero, $\Delta v_j = 0$ for some $j \in R^g$, for example. Since $d_i = 0$, the Δv_i, $i \in R^g$, satisfy a system of the form (12). It follows that $\Delta v_i = a$ for all $i \in R^g$ and, since $\Delta v_j = 0, j \in R^g$, that $a = 0$ and thus $v(i, g) = v(i, f)$ for all $i \in R^g$.

If Δv_T and d_T are the subvectors of Δv and d with the components $i \in T^g$, then (22) is reduced to

$$A \Delta v_T = d_T. \tag{24}$$

$A = I - P_T(g)$ where I is the $m \times m$ identity matrix and $P_T(g) = (p_{ij}^{g(i)})$, $i, j \in T^g$.

The series

$$\sum_{n=0}^{\infty} P_T^n(g) \tag{25}$$

converges according to Theorem 3.2.2 and is equal to $(I - P_T(g))^{-1} = A^{-1}$, as we can see by multiplying the series by $(I - P_T(g))$. A^{-1} thus has only *non-negative* elements. Since $d_i \leqslant 0$ for all $i \in T^g$, it follows that $\Delta v_T = A^{-1} d_T \leqslant 0$ and $v(i, g) \geqslant v(i, f)$ for all $i \in T^g$. All column vectors of A^{-1} have to differ from zero: otherwise A^{-1} would be singular. By assumption we have at least one $d_j < 0$ for some $j \in T^g$. Hence there exists at least one $i \in T^g$ for which $\Delta v_i < 0$ or $v(i, g) > v(i, f)$. The theorem is proved. ∎

According to Theorem 4 one of the following three cases will appear in the policy iteration method: (i) $g = f$ (all $G(i, f)$ empty), (ii) $g \neq f$ and $x(g) > x(f)$, (iii) $g \neq f, x(g) = x(f), v(i, g) \geqslant v(i, f)$ for all $i = 1, \ldots, N$, but $v(i, g) > v(i, f)$ for at least one $i \in T^g$. In case (i) the method terminates with a policy $f = g$ which, by Theorem 3, is optimal. Since there are only finitely many elements $g \in F$, the cases (ii) and (iii) can only occur finitely often. The policy iteration method therefore has to terminate after finitely many steps. Summarizing we have the following result:

Corollary

The policy iteration method terminates with a policy g after finitely many steps. For all policies $f \in F$ we have either $x(f) < x(g)$ or $x(f) = x(g)$ and $v(i, f) \leqslant v(i, g)$ for $i = 1, 2, \ldots, N$, if the relative values are chosen in such a way that $v(j, f) = v(j, g) = 0$ for some $j \in R^g$.

If we minimize instead of maximize, then the directions of the inequalities have to be reversed, and, particularly in (15) and (19), max has to be replaced by min.

Examples

(a) *Computational example.* Let us once again return to Example (5.1.c). This time let the average cost per period be minimized (no discounting). For the policy iteration we shall start from the optimal policy $f(1) = 2, f(2) = 1$ of the problem with discounting (Example (5.2.b)). Let $v_2 = 0$. Then the system (10) for the determination of value reads

$$x + v_1 = 0 + 0.6\,v_1,$$
$$x \quad\quad = 20 + 0.7\,v_1.$$

These equations have the solution $x = 7.27$, $v_1 = -18.18$. For the policy improvement we compute $x + v_1 = -10.91$ and $x + v_2 = x = 7.27$. The left hand sides of (15) become for state 1 and action 1 and state 2 and action 2, respectively:

$$5 + 0.9 \times (-18.18) = -11.36 < -10.91,$$
$$15 + 0.2 \times (-18.18) = 11.36 > 7.27.$$

The new policy $f(1) = 1, f(2) = 1$ thus has to be better than the old one.

For the new policy, (10) reads, again with $v_2 = 0$,

$$x + v_1 = \quad 5 + 0.9\,v_1,$$
$$x \quad\quad = 20 + 0.7\,v_1,$$

from which it follows that $x = 6.875$ and $v_1 = -18.75$.

The right hand sides of (15) become equal to $x + v_1 = -11.88$ and $x + v_2 = x = 6.875$, respectively. The left hand sides of (15) for state 1 and action 2 and state 2 and action 2 are

$$0.6 \times (-18.75) = -11.25 > -11.88,$$
$$15 + 0.2 \times (-18.75) = 11.25 > 6.875.$$

Thus no further policy improvement is possible and the policy $f(1) = f(2) = 1$ is optimal.   □

(*b*) *Replacement problems.* Depending on the circumstances it may be useful to minimize the average cost per period in the replacement problem, Example (5.1.*b*). The reader must note, however, that for this example the Markov chain may have more than one recurrent class for a possible policy! The assumptions of this section are not satisfied. If one considers, however, the class of special policies introduced in Example (5.2.*c*), where either $f(i) = n$ or $f(i) = m$ if $f(i) \neq n$, one can see that the Markov chain has only one recurrent class (and additional transient states) for such policies. If we apply the policy iteration method to such an initial policy f regardless of the above remarks, the policy improvement can be done in such a way that it determines the minimum of

$$B_i + p_i v(i + 1, f) + (1 - p_i)v(N, f)$$

and

$$\min_{k=1,\ldots,N-1} \{C_k - T_i + B_k + p_k v(k + 1, f) + (1 - p_k)v(N, f)\}.$$

In this case the new policy g will have the *same* special form (see Example (5.2.*c*)) and the whole of the policy iteration method moves within this class of special policies. The theory developed in this section can be applied to this class of policies. This means that we can find with the policy iteration method the policy g with the smallest average cost per period among all the policies of the same class. Furthermore, N is always recurrent. If we put $v(N, f) = 0$ for all f of this class, then $v(i, f) \geqslant v(i, g)$ for all f of the class ($\geqslant$ since we deal here with a minimization) and for all i which are transient under g. $-v(i, f)$ can be considered as saving if one can start with a machine of age i instead of a broken down machine (state N). The policy iteration method thus not only minimizes the average cost but also maximizes the savings for all transient states under the optimal policy g. $\square$

(*c*) *Terminating processes.* Let $N + 1$ be an *absorbing* state and let the states $i = 1, \ldots, N$ be *transient* for all policies. Furthermore, let $r_{N+1}^{f(N+1)} = 0$ for all $f \in F$. This implies that $x(f) = 0$ for all $f \in F$. The maximization or minimization of $x(f)$ thus becomes meaningless. If we put $v(N + 1, f) = 0$ we have $F_f^h(i) \rightarrow y(i, f) = v(i, f)$ as $h \rightarrow \infty$ (see (4), (6)); the relative values are equal to the bias values which in turn are equal to the expected pay-offs up to absorption. It may be a meaningful problem to maximize or minimize the expected pay-offs up to absorption. According to the corollary of Theorem 4 the policy iteration method does just that.

If, for example, $r_i^{f(i)} = 1$ for all $i = 1, \ldots, N$ and all $f \in F$, then $y(i, f)$ is the expected number of periods up to absorption. The minimization of $y(i, f)$ thus corresponds to a problem of *time-optimal* (shortest) transition of a system into a certain state $N + 1$. If, furthermore, r_{ij} are 'distances' between the two states i and j and if we put

$$r_i^{f(i)} = \sum_{j=1}^{N+1} p_{ij}^{f(i)} r_{ij},$$

then the minimization of $y(i, f)$ corresponds to the minimization of the expected path length from i to $N + 1$. We thus have a stochastic variant of the *problem of the shortest path length*.

Since $x(f) = 0$, (20) takes the form

$$v_i = r_i^{f(i)} + \sum_{j=1}^{N} p_{ij}^{f(i)} v_j, \quad i = 1, \ldots, N,$$

if $v_{N+1} = 0$. In the policy improvement (see (15)) we consider the inequalities

$$r_i^a + \sum_{j=1}^{N} p_{ij}^a v(j, f) > v(i, f), \quad a \in A_i.$$

A comparison of these relations with the corresponding relations (5.2.8) and (5.2.6) for the case *with* discounting shows considerable *formal* agreement.

If we put in (5.2.8) and (5.2.6) $\beta p_{ij}^a = q_{ij}^a$, then (5.2.8) and (5.2.6) become precisely relations of the above kind. The infinite-step Markov decision problem with discounting thus corresponds to a problem with terminating processes without discounting, with a fictitious absorbing state

$N + 1$, where we introduce the transition probabilities $q^f_{iN+1} = 1 - \beta(p^f_{i1} + \ldots + p^f_{iN}) > 0$ and $q^f_{ij} = \beta p^f_{ij}, i, j = 1, \ldots, N$. The discounted expected values of the pay-offs are thus equal to the undiscounted pay-offs up to absorption.

Finally we should like to point out that the policy iteration method may terminate with *different* policies. All these policies have the *same* profit rate. On the other hand they may have different *bias values*. One can try in addition to optimize the bias values, but this would go beyond the scope of this presentation.

Example

(*d*) *Optimization of bias values.* Let $I = \{1, 2\}, A_1 = \{1, 2\}, A_2 = \{1\}$. Therefore there exist two policies, $f = (1, 1)$ and $g = (2, 1)$. Let $p^1_{11} = p^1_{12} = 0.5, p^2_{11} = 0, p^2_{12} = 1$ and $p^1_{21} = p^1_{22} = 0.5$. Let the pay-offs be $r^1_1 = 3, r^2_1 = 6$ and $r^1_2 = -3$. For both policies f and g we obtain the profit rate $x = 0$ and the relative values $v_1 = 6$ and $v_2 = 0$. For both policies f and g, $G(1, f)$ and $G(1, g)$, respectively, are empty and the policy iteration method can thus terminate with g as well as with f. The stationary distributions are, however, different for f and g, i.e. $(1/2, 1/2)$ and $(1/3, 2/3)$, respectively. The bias values, defined by (6), differ for f and g and are $(3, -3)$ and $(4, -2)$, respectively. Thus g has *greater* bias values than f and is therefore preferable for maximization. $\square$

Notes to Chapter 5

The principle of optimality is due to *R. E. Bellman* [2]. [6] contains a very general treatment of the foundations of dynamic programming . The deterministic sequential decision problem as introduced in Example (5.1.*d*) is of great practical importance. [12], [13], [14], [15] extensively discuss deterministic dynamic programming.

The Markovian decision problem, especially the infinite step problems were first treated in a systematic manner by *R. A. Howard* [7]. The subject is also dealt with extensively in [4] and [11]. Policy iteration is closely related to linear programming; see [11].

The Markovian decision problem, as introduced here, can be extended in several directions. Non-denumerable state spaces may be introduced, see notes to Chapter 3. Such more general problems are discussed in [6] and [11]. Furthermore it is also possible to formulate Markovian decision problems with continuous time parameter and to study optimal control of diffusion processes; see [5] and [9]. Moreover Markov renewal processes can be optimized in a similar manner (see Chapter 7 and [8], [11]). Finally we refer to dynamic stochastic games where the process is controlled simultaneously by two or more conflicting decision makers; see [11].

For applications of dynamic programming, see [1]. Important areas of applications are problems of maintenance, replacement, investment and finance in general, and inventory. The optimal control of queues is another important and interesting field of application; but in general the state space will be infinite and the basic theory has to be extended to cover this case. Finally, ideas of dynamic programming are applied to analyse decision trees; see [3] and [10].

6

Simulation and Monte-Carlo methods

6.1. Random numbers

In probability theory *simulation* means the *numerical construction of samples of stochastic processes.* This is of great interest since the consideration of samples gives a better understanding of stochastic processes whose analytic examination is difficult or even impossible. The situation can be compared with the numerical solution of ordinary differential equations which often appears as the only possibility of analysing a given system in detail. Almost arbitrarily complex processes can be simulated so that simulation is a universally applicable instrument for investigation − but one which, however, again has its own peculiar limitations.

In connection with statistical methods simulation allows the numerical solution of problems in probability theory and other areas. The tools developed in this connection are called *Monte-Carlo methods.* This striking term derives from the fact that for these methods roulette, or an analogue of it, plays a fundamental role in the generation of random events.

According to the introductory definition, simulation is the imitation of events in prescribed stochastic experiments. Its basis is a special *canonical* experiment. This experiment is defined by a sequence $X_1, X_2, X_3, \ldots$ of *independent* random variables which all have *uniform distribution* in the interval $[0, 1]$. This uniform distribution is defined by the following distribution function $H(x)$ and density $h(x)$ (see Example $(1.2.e)$):

$$H(x) = \begin{cases} 0, & \text{for } x < 0, \\ x, & \text{for } 0 \leqslant x \leqslant 1, \\ 1, & \text{for } x > 1, \end{cases} \qquad h(x) = \begin{cases} 0, & \text{for } x < 0, \\ 1, & \text{for } 0 \leqslant x \leqslant 1, \quad (1) \\ 0, & \text{for } x > 1. \end{cases}$$

The associated expected value μ_H and variance σ_H^2 are (see Example $(1.3.f)$):

$$\mu_H = \int_0^1 x\,\mathrm{d}x = \frac{1}{2}; \quad \sigma_H^2 = \int_0^1 (x - \tfrac{1}{2})^2\,\mathrm{d}x = \frac{1}{12}. \quad (2)$$

Numeric sequences $\{x_1, x_2, x_3, \ldots\}$, which can be considered as samples or realizations of $\{X_1, X_2, X_3, \ldots\}$, are called *canonical random numbers.*

A first task of the simulation technique consists of making canonical random numbers available. One can try, for example, to construct some physical experiments which are believed to correspond to the canonical experiment. Or one can collect statistics from random experiments as used for lottery or roulette. Since, however, simulation computations take on such a complexity that they require the use of a computer, it has proved more advantageous to produce random numbers by *computation*; and this in spite of the obvious incompatibility between the necessarily deterministic computational procedure and randomness. In these computations one is satisfied with the fact that the resulting random numbers appear to be random without really being random. We speak of *pseudo-random numbers*.

Computation is only possible with a bounded number of digits, so that the continuous uniform distribution must be approximated by a discrete one. Under these conditions we can, furthermore, work with bounded integers $\{x'_1, x'_2, x'_3, \ldots\}$ which can easily be normalized to numbers between 0 and 1 by a suitable division. The simplest generating procedure is given by *recursion formulae* of the form $x'_{i+1} = f(x'_i)$. Among the many suggestions the following recursive relation has so far proved to be the most suitable:

If x'_i is given as an integer, then x'_{i+1} is the integral remainder of the division ax'_i/m, where a, m are given integers with $a < m$.

We have thus $ax'_i = qm + x'_{i+1}$, where q is the greatest integer such that $qm \leqslant ax'_i$. For this we write in short

$$x'_{i+1} \equiv ax'_i \quad (\text{mod } m) \tag{3}$$

(in words: x'_{i+1} is congruent to ax'_i modulo m). Readers familiar with number theory will recognize the connection with residue classes and congruence classes of numbers. $x \equiv y \,(\text{mod } m)$ means that $x - y = 0$ or is divisible by m. From this follow in particular the following important rules: if $u \equiv v\,(\text{mod } m)$ and $x \equiv y\,(\text{mod } m)$ we also have $u \pm x \equiv v \pm y\,(\text{mod } m)$ and $ux \equiv vy\,(\text{mod } m)$. Thus we obtain from $x_1 \equiv ax_0\,(\text{mod } m)$ also $ax_1 \equiv a^2 x_0\,(\text{mod } m)$ and by (3) it follows that $x_2 \equiv a^2 x_0\,(\text{mod } m)$ and finally, in general, $x_n \equiv a^n x_0\,(\text{mod } m)$.

For an integral initial value or *seed* x'_0 the $x'_i, i = 1, 2, \ldots$, are, according to (3), integers with the possible values $0, 1, 2, \ldots, m - 1$. We obtain thus numbers x_i between 0 and 1 if we divide the x'_i by m:

$$x_i = x'_i/m. \tag{4}$$

The generation of random numbers by (3), (4) is called the *multiplicative congruence method*. Generalization from (3) to $x'_{i+1} \equiv ax'_i + c\,(\text{mod } m)$ has also been considered (mixed congruence method). We will however limit ourselves to a consideration of multiplicative congruential random numbers as most available computer programs are based on this method.

Do the integers defined by (3) appear to be sufficiently random? This depends on the choice of the *factor a*, the *module m* and of the *seed x'_0*.

Since the numbers of the sequence $x'_i = 0, 1, 2, \ldots$ can only take m different values, the sequence has to repeat itself after m steps at the latest. An integer $x' \in M = \{0, 1, 2, \ldots, m - 1\}$ *leads to* $y' \in M, x' \Rightarrow y'$, if y' appears as a term in sequence (3) with seed x', i.e. if there exists an $n > 0$ such that $y' = a^n x'$ (mod m). Terms x' and y' *communicate*, $x \Leftrightarrow y'$, if $x' \Rightarrow y'$ and $y' \Rightarrow x'$. If $x' \Rightarrow x'$ we call x' *recurrent*, and the smallest number $n > 0$ for which $x' \equiv a^n x'$ (mod m), is called the *period* of x'. If x' does not return to itself, we call x' *transient*. A transient number can only occur *once* in every sequence (3) and cannot communicate with another number. The relation $\Leftrightarrow$ is therefore an equivalence relation on the set of the recurrent numbers which it decomposes into disjoint equivalence classes of communicating numbers. If x'_0 is a recurrent number, then the sequence x'_i with seed x'_0 obviously runs *exactly once* through all numbers of the equivalence class of x'_0, before returning again to the value x'_0; hence all numbers of the same class have the *same* period which in turn is equal to the number of the elements of the class. From every seed x'_0 the sequence (3) reaches an equivalence class since at some stage, as has already been mentioned, a value must re-occur. In order to maintain the appearance of randomness the period of the class should at least be large enough so that it is not completely exhausted when used in practice. For simulation we often need thousands of random numbers. Since the period is bounded by the module m, m must always be chosen sufficiently large. But this is not all. If a and m have a *common factor* $d > 1$, then also all the x'_i, $i = 1, 2, \ldots$, of the sequence (3) are multiples of d. All recurrent classes thus have periods which are smaller than m/d. For obvious reasons one should therefore choose a and m *relatively prime*. We then have

Theorem 1
If a and m are relatively prime, then every $x' \in M = \{0, 1, 2, \ldots, m - 1\}$ is *recurrent*.

Proof. $x'_{n+1} \equiv ax'_n \pmod{m}$ has, under the circumstances, a unique solution x'_n: if we also have $x'_{n+1} \equiv ax''_n \pmod{m}$, then it follows that $0 \equiv a(x'_n - x''_n) \pmod{m}$ and hence that $x'_n \equiv x''_n \pmod{m}$, if a is relatively prime to m. If, therefore, $x'_0 = x' \in M$ and x'_k is the first recurrent number in the sequence x'_i with seed x', then there exists an n such that $x'_k = x'_{k+n}$, and, if $k > 0$ we have, according to the above, also $x'_{k-1} = x'_{k-1+n}$. Contrary to our assumptions x'_k is thus not the first recurrent number and therefore $k = 0$ and x' must be recurrent. ∎

In the following we always assume a and m to be relatively prime. x'_0 and m should also be *relatively prime* as otherwise the x'_i would again be multiples of the common divisor. Conversely, if x'_0 and m are relatively prime, then all x'_i are relatively prime to m and therefore the periods are, at most, equal to the number of the relatively prime numbers to m in $M = \{0, 1, 2, \ldots, m - 1\}$.

The greatest possible length of period of $m - 1$ can thus only be achieved if m is a prime number. A large prime number could be a suitable choice for m. In any case m should obviously not have too many distinct prime factors. For practical reasons m is often chosen to be a power of 2, $m = 2^r$. This is connected with the *binary representation* of numbers in the computer. In integral arithmetic with word length $r + 1$ one binary digit, a so-called *bit*, is reserved for the sign so that $2^r - 1$ is the largest number which can be represented. If we obtain numbers with more than r bits during an arithmetic operation, then the highest binary digits are cut off, i.e. the computer works basically modulo $m = 2^r$; but the digit $r + 1$, the sign, has to be treated differently.

The maximal obtainable length of period in this case is described in the following theorem.

Theorem 2

If $m = 2^r, r \geqslant 3$, then there exist recurrence classes with the maximal period 2^{r-2}. This period is obtained if x_0' is *odd* and $a \equiv 3$ or $a \equiv 5$ (mod 8).

To prove this theorem two lemmas are needed.

Lemma 1

Let $f > 1$ be an integer and let

Then

$$a \equiv \pm 1 \,(\text{mod } 2^f) \quad \text{and} \quad a \not\equiv \pm 1 \,(\text{mod } 2^{f+1}). \tag{5}$$

$$a^2 \equiv 1 \,(\text{mod } 2^{f+1}) \quad \text{and} \quad a^2 \not\equiv 1 \,(\text{mod } 2^{f+2}). \tag{6}$$

Proof. By assumption (5) we have $a = \pm 1 + q2^f$, q *odd*, and hence

$$a^2 = 1 + q2^{f+1}(\pm 1 + q2^{f-1}).$$

Since $q(\pm 1 + q2^{f-1})$ is odd we conclude (6). ∎

Lemma 2

If $a > 1$ and odd, then there exists a *unique* integer $f > 1$ such that

$$a = \pm 1 + 2^f + p2^{f+1} \quad \text{or} \quad a \equiv 2^f \pm 1 \,(\text{mod } 2^{f+1}). \tag{7}$$

If furthermore $1 < a < 2^b - 1$ for some $b \geqslant 3$, then the smallest number d for which $a^d \equiv 1 \,(\text{mod } 2^b)$ is equal to 2^{b-f}.

Proof. If a is odd, then either $a + 1$ or $a - 1$ is a multiple of 4. This means that $a \pm 1 = q2^f$ where q is an odd number, and $f > 1$ is therefore uniquely defined. Let $q = 2p + 1$. We obtain $a = \pm 1 + 2^f + p2^{f+1}$ and hence (7).

If we consider the representation of a in (7), then $f < b$ since $a < 2^b$ and it follows from (7) that $a \equiv \pm 1 \,(\text{mod } 2^f), a \not\equiv \pm 1 \,(\text{mod } 2^{f+1})$. By repeated application of Lemma 1 we obtain $a^2 \equiv 1 \pmod{2^{f+1}}$, $a^2 \not\equiv 1 \pmod{2^{f+2}}$, $a^4 \equiv 1 \pmod{2^{f+2}}$, $a^4 \not\equiv 1 \pmod{2^{f+3}}$, ..., and finally

$$a^{2^{b-f-1}} \not\equiv 1 \pmod{2^b}, \quad a^{2^{b-f}} \equiv 1 \pmod{2^b}. \tag{8}$$

If $d < 2^{b-f}$, then we could write $2^{b-f} = qd + r, r < d$. However, since $a^r \not\equiv 1$ $\pmod{2^b}$, it follows that $r = 0$ and d must be a divisor of 2^{b-f}, i.e. d must be of the form 2^k, $k < b - f$. This, however, is impossible according to (8). Therefore $d = 2^{b-f}$ and the lemma is proved. ∎

Proof of Theorem 2. We have $x'_n \equiv a^n x'_0 \pmod{2^r}$. If a is even, $a = k2^s$, $s < r$, then the period becomes equal to 1; after an initial transient phase $a^n x'_0 \equiv 0 \pmod{2^r}$ and this occurs at the latest for $n \geqslant r/s$. In order to obtain the maximum length of period, a has to be chosen *odd*. According to Theorem 1 the period is equal to the smallest integer d for which $x'_0 \equiv a^d x'_0$ $\pmod{2^r}$. If $x'_0 = k2^s$, k odd, $s \geqslant 0$, then this condition is equivalent to $a^d \equiv 1 \pmod{2^{r-s}}$. If $b = r - s < 3$, then the period is at the most 2. If $b = r - s \geqslant 3$, Lemma 2 can be applied. For $a = 1$ and $a = 2^b - 1$ we obtain the periods 1. For $1 < a < 2^b - 1$ this period becomes, according to Lemma 2, equal to 2^{b-f}, $f > 1$.

For a given b we thus obtain the greatest period with $f = 2$ or $a \equiv 4 \pm 1$ $\pmod 8$, hence $a \equiv 3$ or $a \equiv 5 \pmod 8$. The period is thus $2^{b-2} = 2^{r-s-2}$. The maximal period is obtained if $s = 0$ or if x'_0 is an odd number. The theorem is proved. ∎

Example

(*a*) *Some random number generators used in practice.* Computing centres or computer companies generally offer complete programs for the generation of random numbers. In Table 1 a few examples have been compiled.

Table 1

Module m	Factor a	Period
2^{31}	$2^{16} + 3 = 65\,539$	$2^{29} = 536\,870\,912$
2^{35}	$5^{13} = 1\,220\,703\,125$	$2^{33} = 8\,589\,934\,592$
$2^{31} - 1^*$	$16\,807$	$2^{31} - 2 = 2\,147\,483\,646$

* Prime number, hence period $m - 1$. □

A large length of period is certainly a necessary condition for a satisfactory random number generator but is by no means a sufficient one. The appearance of randomness can only be guaranteed if the statistics of the sequence

$x_1, x_2, \ldots$ agree sufficiently well with the theoretical expectations of the canonical experiment. The degree of agreement depends for many important statistics crucially on the choice of the factor a within the restrictions of Theorem 2. Therefore we have to examine carefully for each choice of the factor a whether a satisfactory behaviour of the sequence of random numbers is obtained.

This can be done by applying a number of *empirical statistical tests* to bounded *subsequences* of the sequence of random numbers. Statistics of the sequence of random numbers can also be considered over a *complete period*. If d is the length of the period we can consider the expected value $\bar{x}$ over the period as well as the variance s^2 over the period

$$\bar{x} = \frac{1}{d} \sum_{i=0}^{d-1} x_i, \quad s^2 = \frac{1}{d} \sum_{i=0}^{d-1} (x_i - \bar{x})^2 \tag{9}$$

which, according to (2), should be approximately equal to $\frac{1}{2}$ and $\frac{1}{12}$, respectively. The *period covariances* r_k,

$$r_k = \frac{1}{d} \sum_{i=0}^{d-1} (x_i - \bar{x})(x_{i+k} - \bar{x}), \quad k = 1, 2, \ldots, \tag{10}$$

should vanish since we obtain from the canonical experiment $E[(X_i - \frac{1}{2})(X_{i+k} - \frac{1}{2})] = 0$ for $k \neq 0$, because of the independence of X_i and X_{i+k}. Also we can consider the distribution of individual random numbers, pairs, triples, etc., over all periods. It is difficult to compute the statistics of random numbers over a complete period since d is, in general, too large in order to evaluate formulae such as (9) and (10) directly numerically. These important questions cannot be pursued here any further and we have to refer the reader to specialized literature. Most computing centres provide tested random number generators so that there is no need for developing one's own generator.

6.2. Simulation of random variables

Canonical random numbers can be used for the construction of samples of quite general stochastic processes. The simplest problem consists of the simulation of random variables Y with a given distribution function $F(y)$. If $X_1, X_2, \ldots$ is a sequence of independent random variables which are uniformly distributed in the interval $[0, 1]$ (canonical experiment, Section 6.1), then the problem can be considered as solved if one succeeds in finding a function f such that the random variable $Y = f(X_1, X_2, \ldots)$ has the given distribution $F(y)$. If we apply the function f to canonical random number $x_1, x_2, \ldots$, then $y = f(x_1, x_2, \ldots)$ can obviously be considered as a sample of the random variable Y with distribution $F(y)$. In the following we shall formulate a few theorems which allow the construction of such transformations of canonical random numbers. These results will form the basis for the simulation of stochastic processes.

Theorem 1

Let $F(y)$ be a *strictly* monotonic continuous distribution function and X a uniformly distributed random variable in $[0, 1]$. Then the random variable

$$Y = F^{-1}(X) \tag{1}$$

has the distribution function $F(y)$.

Proof. Under the assumptions of the theorem the inverse function F^{-1} of F exists. We have

$$P[Y \leqslant y] = P[F^{-1}(X) \leqslant y] = P[X \leqslant F(y)].$$

However, since X is uniformly distributed in $[0, 1]$, we have $P[X \leqslant F(y)] = F(y)$ and hence $P[Y \leqslant y] = F(y)$. ∎

According to Theorem 1 we are thus able to define samples y for a distribution function which satisfies the assumptions of Theorem 1 by solving the equation

$$x = F(y) \quad (x \text{ canonical random number}). \tag{2}$$

This is called an *inversion method*. Equation (2) can, however, only be solved explicitly in special cases (see Examples (a) and (b) below). Otherwise, we have to apply a numerical approximation method for the solution of (2).

Theorem 2

Let $F(y)$ be a *discrete* distribution function which has jumps of heights $p_1, p_2, \ldots$ at $y_1 \leqslant y_2 \leqslant \ldots$ and let X be a uniformly distributed random variable in $[0, 1]$. The random variable defined by

$$Y = y_i, \text{ if } \sum_{j=1}^{i-1} p_j < X \leqslant \sum_{j=1}^{i} p_j, \quad i = 1, 2, \ldots, \tag{3}$$

has the discrete distribution $F(y)$.

Proof. We have $Y = y_i$ if X falls in the interval of length p_i defined by (3). The probability for this is equal to p_i since X is uniformly distributed in $[0, 1]$. Thus we obtain $P[Y = y_i] = p_i$. ∎

From Theorem 2 we obtain a simple method for the simulation of discrete random variables which do not have too many possible values. Geometrically speaking we subdivide the interval $[0, 1]$ into disjoint subintervals of length

$p_i, i = 1, 2, \ldots$. If the canonical number x falls into the ith interval we put $y = y_i$. We are thus able to consider y as the sample for the given discrete distribution. This is called an *interval test method*. Here the order of the subintervals is obviously unimportant.

Theorems 1 and 2 are special cases of the following general theorem.

Theorem 3

If X is a uniformly distributed random variable in $[0, 1]$ and $F(y)$ an *arbitrary* distribution function, then the random variable defined by

$$Y = \inf\{z : X \leqslant F(z)\} \tag{4}$$

has the distribution function $F(y)$.

Proof. Since

$$P[Y \leqslant y] = P[\inf\{z : X \leqslant F(z)\} \leqslant y],$$

we have to prove that this is equal to $F(y)$. If we have $X \leqslant F(y)$, then it follows from the monotonicity of $F(y)$ that $\inf\{z : X \leqslant F(z)\} \leqslant y$, and conversely, if $X > F(y)$ it follows that $\inf\{z : X \leqslant F(z)\} > y$ for reasons of monotonicity and since $F(y)$ is continuous from the right. Thus the two events $[\inf\{z : X \leqslant F(z)\} \leqslant y]$ and $[X \leqslant F(y)]$ are identical and we have

$$P[Y \leqslant y] = P[X \leqslant F(y)] = F(y).$$

The second equality holds since X is uniformly distributed in $[0, 1]$. Thus the theorem is proved. ∎

The following theorem sometimes leads to simple numerical methods.

Theorem 4

Let Z_1 be a random variable with distribution function $F(x)$, Z_2 a random variable with a continuous distribution function $G(x)$ which is independent of Z_1, and $r(x)$ a continuous function. If

$$\frac{1}{c} = \int_{-\infty}^{\infty} G(r(x)) \, dF(x) \neq 0, \tag{5}$$

then the conditional distribution of Z_1 under the condition $Z_2 \leqslant r(Z_1)$ is

$$P[Z_1 \leqslant z | Z_2 \leqslant r(Z_1)] = c \int_{-\infty}^{z} G(r(x)) \, dF(x). \tag{6}$$

Proof. From the definition (1.1.10) of conditional probabilities we have

$$P[Z_1 \leqslant z | Z_2 \leqslant r(Z_1)] = \frac{P[Z_1 \leqslant z; Z_2 \leqslant r(Z_1)]}{P[Z_2 \leqslant r(Z_1)]}. \tag{7}$$

However,

$$P[x - \mathrm{d}x \leqslant Z_1 \leqslant x, Z_2 \leqslant r(x)] = G(r(x))\mathrm{d}F(x). \tag{8}$$

We thus obtain the numerator of (7) by integration of (8) from $-\infty$ to z and the denominator of (7) by integration of (8) from $-\infty$ to $+\infty$ ∎

If $F(x)$ has a density function $f(x)$ then $P[Z_1 \leqslant z | Z_2 \leqslant r(Z_1)]$ has the density function $h(x) = cf(x)G(r(x))$. When applying Theorem 4, $G(x)$ is mostly assumed to be a uniform distribution in the interval $[0, 1]$. If, in addition, we have $0 \leqslant r(x) \leqslant 1$ for all x for which $f(x) \neq 0$, then it follows that

$$h(x) = \frac{f(x)r(x)}{\int_{-\infty}^{\infty} f(x)r(x)\mathrm{d}x}. \tag{9}$$

If a density function $h(x)$ is *given*, then Theorem 4 or (9) may be used in order to simulate a random variable Y with density $h(x)$, respectively: suppose that random variables with density $f(x)$ can be simulated (for example, by the inversion method) and suppose we have $f(x) \neq 0$ for all x with $h(x) \neq 0$. The function $r(x) = kh(x)/f(x)$ (if $h(x) > 0$), $r(x) = 0$ (if $h(x) = 0$) satisfies equation (9) for the given density $h(x)$. The constant k is to be chosen in such a way that $0 < r(x) \leqslant 1$. This is possible if $\sup(h(x)/f(x)) = M < \infty$ (the supremum is taken over all x for which $f(x) \neq 0$). Then we may choose k in the interval $0 < k \leqslant 1/M$. If Z_1 is distributed according to $f(x)$ and $Z_2 = X$ is uniformly distributed in $[0, 1]$, then Z_1 is distributed with density $h(x)$ under the condition $X \leqslant r(Z_1)$ according to Theorem 4 and (9). Samples with density $h(x)$ are thus constructed as follows: we find a sample z of the random variables Z_1 distributed according to $f(x)$ and a sample x of the uniformly distributed random variable X. If we have $x > r(z) = kh(z)/f(z)$, we discard the pair of samples and find a new pair of random numbers z, x etc. until we obtain $x \leqslant kh(z)/f(z)$. We may then consider z as sample for the density $h(x)$. This method is called the *rejection method*.

The probability $P[X \leqslant r(Z_1)]$ represents the expected percentage of the sample pairs which do not have to be discarded. We have

$$P[X \leqslant r(Z_1)] = \int_{-\infty}^{\infty} r(x)f(x)\mathrm{d}x = k.$$

One tries, of course, to keep this percentage as large as possible since it represents the ratio between the number of used and the number of generated random numbers and thus constitutes a measure for the *effectiveness* of the method. It is thus advisable to choose k as near to $1/M$ as possible.

If $h(x)$ vanishes outside a *finite* interval $[a, b]$ we can take for $f(x)$ the density of the uniform distribution in the interval $[a, b]$, $f(x) = 1/(b - a)$ for $a \leqslant x \leqslant b$, $f(x) = 0$ otherwise. It is very easy to determine samples for such a uniform distribution (cf. Example (*a*) below).

We can also develop a rejection method for *discrete* distributions. Let $F(x)$

be the distribution function of a discrete distribution for which the m possible values $z_1, z_2, \ldots, z_m$ are all assumed with the same probability $1/m$. If $p_1, p_2, \ldots, p_m$ are probabilities, where $p_1 + p_2 + \ldots + p_m = 1$, then we choose $r(x)$ such that $r(z_i) = p_i$ and, if $G(x)$ is the uniform distribution in the interval $[0, \max p_i]$, then Z_1, according to (6), and under the condition $Z_2 \leqslant r(Z_1)$, takes just the values $z_1, z_2, \ldots, z_m$ with the probabilities $p_1, p_2, \ldots, p_m$. Indeed, we have under these assumptions $G(x) = x/\max_i p_i$, $c = m \times \max_i p_i$ and thus $P[Z_1 = z_1 | Z_2 \leqslant p_i] = p_i] = p_i$. The rejection method thus goes as follows: we determine a sample z of Z_1 with the discrete distribution $F(x)$ (see Example (a) below) and a sample u according to the uniform distribution in $[0, \max p_i]$. If $z = z_i$, we check the inequality $u \leqslant p_i$. If it is satisfied we may consider $z = z_i$ as a sample for the discrete distribution $P[Y = z_i] = p_i, i = 1, \ldots, m$; otherwise we determine a new pair of samples z and u and repeat the computation.

Inversion and rejection methods may be combined. In the rejection method, samples of $F(x)$ may be generated by the inversion method, or alternatively, also be generated by the rejection method. They may, however, also be inter-connected by the so-called *composition method*. Let $G(y)$ be an arbitrary distribution function, and $f(x, y)$ a density for all y. Then

$$h(x) = \int_{-\infty}^{\infty} f(x, y) \, dG(y) \tag{10}$$

is a density function. In order to construct a sample z for $h(x)$, we first determine a sample $\hat{y}$ for $G(y)$ and afterwards a sample z for the density $f(x, \hat{y})$. The composition method is of particular interest if the density $h(x)$, about to be simulated, can be represented in the form $h(x) = p_1 f_1(x) + p_2 f_2(x)$, where $p_1 + p_2 = 1, p_1$ is much greater than p_2 and $f_1(x)$ can easily be simulated. In the majority of cases $\hat{y} = 1$, and the sample z for $h(x)$ is obtained by the easy simulation of the density $f_1(x)$. The difficult density $f_2(x)$ has only to be simulated in relatively few cases.

Examples

In the following we shall consider a few common *special* distributions. In some cases we can apply the general methods constructed above and in the other cases one can develop special methods.

(a) *Uniform distributions* (see Example (1.2.e)). For the distribution function $F(y) = 0$ for $y < a, F(y) = (y - a)/(b - a)$ for $a \leqslant y \leqslant b, F(y) = 1$ for $y > b$, we have the inverse $F^{-1}(x) = a + (b - a)x$ in the interval $[a, b]$. If x is a canonical random number, then, according to the *inversion method*, $y = a + (b - a)x$ is a sample of the uniform distribution in the interval $[a, b]$.

The uniform distributions are particularly important for the rejection method. If this is applied to the density $f(x)$ of a uniform distribution in the interval $[a, b]$, we can specialize the method as follows: we choose

$r(z) = k(b - a)h(z), a \leqslant z \leqslant b$ and $0 < k \leqslant 1/M = 1/\sup((b - a)h(z))$. If x_1, x_2 are a pair of canonical random numbers, then these are rejected if $x_2 > r(a + (b - a)x_1) = k(b - a)h(a + (b - a)x_1)$ and we consider a new pair, etc. until we get $x_2 \leqslant k(b - a)h(a + (b - a)x_1)$. In this case we can consider $y = a + (b - a)x_1$ as a sample for the density $h(x)$. This special rejection method can be applied if the density $h(x)$ to be simulated is bounded and vanishes outside the interval $[a, b]$.

A *discrete uniform* distribution whose values $i = 1, 2, \ldots, m$ all have the same probability $1/m$, can be simulated either by the interval test method or, much more simply, by constructing a canonical random number x, taking the product mx and by *rounding it up* to the next integer. A discrete distribution for which the m values $z_1, z_2, \ldots, z_m$ are all taken with the same probability $1/m$, can be simulated by choosing an index $i = 1, \ldots, m$, according to the discrete uniform distribution by the method just described, and then taking the corresponding value z_i as sample. This can be applied to the simulation of discrete distributions by the rejection method.   □

(b) *Exponential distributions* (see Example (1.2.f)). For the exponential distribution $F(y) = 0, y < 0, F(y) = 1 - e^{-\lambda y}, y \geqslant 0$, the equation (2) of the inversion method reads $x = 1 - e^{-\lambda y}$, from which it follows that $y = -(1/\lambda)\log(1 - x)$. Since $1 - x$ is, like x, uniformly distributed in $[0, 1]$ we can just as well take $y = -(1/\lambda)\log x$ as a sample for the exponential distribution.   □

(c) *Normal distributions* (see Example (1.2.g)). For the simulation of normal distributions neither the inversion method nor the rejection method are particularly suitable. However, there exist a number of special methods. By the central limit theorem (1.4.22) the random variable

$$Y = \frac{\sum_{i=1}^{n} X_i - \frac{1}{2}n}{\sqrt{(\frac{1}{12}n)}}$$

is, for sufficiently large n, approximately normally distributed with expected value 0 and variance 1, if the X_i are independent and uniformly distributed in the interval $[0, 1]$ (with the expected value $\frac{1}{2}$ and variance $\frac{1}{12}$, see (6.1.2)). For $n = 12$ the approximation can be considered as sufficient for most purposes. This choice of n has the advantage that the square root disappears in the denominator of Y. If $x_1, \ldots, x_{12}$ are canonical random numbers, we may consider

$$y = \sum_{i=1}^{12} x_i - 6$$

as the sample of a $N(0, 1)$ normal distribution.

This method, based on the central limit theorem, has the disadvantage that the extreme values of the normal distribution are represented relatively badly; for $n = 12$, for instance, we have, according to the above formula, in any case

$-6 \leqslant y \leqslant 6$. This disadvantage can be avoided by another method. Let Y_1 and Y_2 be two *independent* normally distributed $N(0, 1)$ random variables with the joint density function

$$f(y_1, y_2) = \frac{1}{2\pi} e^{-(y_1^2 + y_2^2)}.$$

The point (Y_1, Y_2) on the two-dimensional plane has the polar coordinates (R, θ). The random variable θ is obviously *uniformly* distributed in the interval $[0, 2\pi]$. In order to determine the distribution of the random variable $R^2 = Y_1^2 + Y_2^2$ we compute the probability $P[R^2 \leqslant r^2] = P[Y_1^2 + Y_2^2 \leqslant r^2]$. This probability is obtained by integrating the density function $f(y_1, y_2)$ over the disc $y_1^2 + y_2^2 \leqslant r^2$. After changing the integration variables into polar coordinates we obtain

$$P[R^2 \leqslant r^2] = \int_0^{2\pi} \frac{d\theta}{2\pi} \int_0^r r e^{-r^2/2} dr = 1 - e^{-r^2/2}.$$

R^2 is thus *exponentially* distributed with parameter $\frac{1}{2}$. Furthermore R^2 and θ are mutually *independent*. If x_1 and x_2 are two canonical random numbers then $z_1^2 = -2 \log x_1$ and $z_2 = 2\pi x_2$ can be considered as samples of the random variables R^2 and θ (see Examples (*b*) and (*a*)). Since we have $Y_1 = R \cos \theta$, $Y_2 = R \sin \theta$,

$$y_1 = (-2 \log x_1)^{1/2} \cos 2\pi x_2,$$

$$y_2 = (-2 \log x_1)^{1/2} \sin 2\pi x_2,$$

are samples of Y_1 and Y_2, i.e. they are independent samples of $N(0, 1)$ distributions.

Thus we have two simple methods for the simulation of $N(0, 1)$ random variables. For the simulation of normally distributed random variables with arbitrary expected value μ and variance σ^2 ($N(\mu, \sigma^2)$ variables) we use the fact that $Z = \sigma Y + \mu$ is an $N(\mu, \sigma^2)$ variable. This follows from

$$P[Z \leqslant z] = P[\sigma Y + \mu \leqslant z] = \frac{1}{\sqrt{(2\pi)}} \int_{-\infty}^{(z-\mu)/\sigma} e^{-x^2/2} dx$$

$$= \frac{1}{\sqrt{(2\pi)}\sigma} \int_{-\infty}^{z} e^{-(y-\mu)^2/2\sigma^2} dy$$

with the change of variable $x = (y - \mu)/\sigma$. Thus if y is a sample of $N(0, 1)$, then $z = \sigma y + \mu$ is a sample of $N(\mu, \sigma^2)$. $\qquad \square$

(*d*) *Binomial distributions* (see Example (1.4.*b*)). The interval test method or rejection method can be applied. If, however, we want to write a general computer program for generating binomially distributed random variables for arbitrary parameters n and p, then these methods are not very suitable, since the probabilities $p_i = \binom{n}{i} p^i (1-p)^{n-i}$ for various n and p can hardly be stored and have to be computed over and over again.

The following method is more suitable: we simulate n independent Bernoulli variables $Y_i, i = 1, \ldots, n$, which take the value 1 with probability p and the value 0 with probability $1 - p$ (see Example (1.2.b)). This can be done by the interval test method. If the x_i are n canonical random numbers we write $y_i = 1$, if $x_i \leqslant p$, and $y_i = 0$, if $x_i > p$. Since $Y_1 + \ldots + Y_n$ is binomially distributed, $z = y_1 + \ldots + y_n$ can be considered to be a sample of the binomial distribution (see Example (1.4.b)).

For large n this method is too elaborate. However, according to the central limit theorem (1.4.25), for large n, $Z = Y_1 + \ldots + Y_n$ is approximately normally distributed with expected value np and variance $np(1-p)$ (see Examples (1.4.b) and (1.4.d)). The approximation is satisfactory if we have neither $np < 5$ nor $n(1-p) < 5$. We can therefore, according to Example (c), determine a sample y of $N(0, 1)$, compute $z = (np(1-p))^{1/2} y + np$ and round to the nearest integer in the interval from 0 to n. The result may be considered as a sample of the binomial distribution.   □

(e) *Geometric distributions* (see Example (1.2.c)). When applying the interval test method we must note that the geometric distribution has *infinitely* many possible values. The sums of the $p_i(3)$ can therefore not all be computed in advance and then stored in the computer. But we can apply the interval test method by computing the $p_j = (1-p)p^i$ *recursively* to such an extent as is necessary for the execution of the test sequence.

Another simple method is based on the simulation of exponentially distributed random variables. If Z is exponentially distributed with parameter λ we have for $i = 0, 1, 2, \ldots$

$$P[i \leqslant Z < i + 1] = e^{-\lambda i}(1 - e^{-\lambda}).$$

This is a geometric distribution with $p = e^{-\lambda}$. For the simulation of a geometric distribution with parameter p we determine a sample z for the exponential distribution with parameter $\lambda = -\log p$ (cf. Example (b)) and round z to the nearest integer. The result is a sample of the geometric distribution with parameter p.   □

(f) *Poisson distributions* (see Example (1.2.d)). As with the geometric distribution in Example (e) the interval test method can be used in a similar way with recursive computation of the probabilities $p_i = (\lambda^i/i!) e^{-\lambda}$. According to Example (2.1.a) we can also simulate a *renewal process* with exponential distribution with parameter λ as renewal distribution (*Poisson process*) and determine how many renewals fall into the interval $(0, 1]$. The number of these renewals has exactly the desired Poisson distribution. One thus generates a series of exponentially distributed samples $z_1, z_2, \ldots$ until the sum

$z_1 + z_2 + \ldots + z_{n+1}$ exceeds 1 for the first time. Then n can be considered as sample value for the Poisson distribution. □

6.3. Simulation of stochastic processes

We shall restrict ourselves to stochastic processes which either have *discrete time parameter* or whose sample functions are *step functions*, i.e. are piecewise constant functions if the processes have continuous time parameter. In this case the function values as well as the jumps of the sample functions can be random. Let such a process be called a *jump process*. The majority of the processes which are simulated in operations research can be associated to one of the two classes of stochastic processes.

Let $\{X_n, n = 0, 1, 2, \ldots\}$ be a *stochastic process with discrete parameter*. The simulation of such a process involves the construction of samples of the n-dimensional vectors $\{X_0, X_1, \ldots, X_{n-1}\}, n = 1, 2, \ldots$. It is only natural to choose the simulation in such a way that a sample $\{x_0, x_1, \ldots, x_{n-1}, x_n\}$ for the $(n + 1)$-dimensional vector $\{X_0, X_1, \ldots, X_n\}$ is obtained *recursively* from a sample $\{x_0, x_1, \ldots, x_{n-1}\}$ for the n-dimensional vector $\{X_0, X_1, \ldots, X_{n-1}\}$. If we know the *conditional* distribution function of X_n, given X_0, $X_1, \ldots, X_{n-1}, F_n(x|x_0, \ldots, x_{n-1})$, then the sample x_n can be determined according to a suitable method of Section 2 for X_n, given $X_0 = x_0, X_1 = x_1, \ldots, X_{n-1} = x_{n-1}$. To start this recursive simulation an initial value x_0 of X_0 must be simulated according to the absolute distribution $F_0(x) = P[X_0 \leqslant x]$; this is also done by the methods of Section 2.

This simulation method is particularly simple for *Markov chains* with *discrete* parameter, since, because of the Markov property (3.1.1), $F_n(x|x_0, \ldots, x_{n-1}) = F_n(x|x_{n-1})$. It is therefore sufficient to know the value i of X_{n-1} in order to simulate the next value X_n. The probabilities of the possible values j of X_n, given $X_{n-1} = i$, are determined by the transition probabilities p_{ij}. A discrete Markov chain can therefore be simulated as follows: the initial state X_0 is simulated for the *initial distribution* $p_i, i = 1, 2, \ldots$. If we obtain in the course of the simulation $X_{n-1} = i$, then the next state X_n is simulated according to the distribution $p_{ij}, j = 1, 2, \ldots$. For the simulation of the occurring discrete distributions we can apply the interval test method of Section 6.2. In many applications the p_{ij} are not given explicitly, but only implicitly defined by probability experiments, whose outcomes have distributions corresponding to the transition probabilities. We then simulate according to the prescribed transition probabilities, if we simulate the defining probability experiments. The following examples will illustrate this.

Examples

(*a*) *Branching processes* (see Example (3.1.*e*)). The determination of the transition probabilities p_{ij} from the distribution of offspring Z can be quite complicated. If, however, the nth generation consists of i individuals then the offspring of each individual z_k, $k = 1, \ldots, i$, can be simulated, according to independent copies of the distribution Z. The $(n + 1)$th generation consists of $j = z_1 + \ldots + z_i$ individuals.   □

(*b*) *Stock-keeping* (see Example (3.1.*f*)). Again it is not necessary to compute the transition probabilities for this example explicitly. If the stock $X_n = i$ at the beginning of the nth day and if we simulate the demand Z of the nth day by the (discrete) demand distribution $P[Z = k]$ with the sample z, then the stock j at the beginning of the $(n + 1)$th day is equal to $i - z$ if $i - z > s$, or equal to S if $i - z \leqslant s$.   □

(*c*) *Renewal processes.* With the same method we can simulate *renewal processes* for a renewal distribution $F(x)$. We determine samples x_n, $n = 1$, $2, \ldots$, for $F(x)$ for the successive renewal intervals. The successive renewal points s_n can easily be obtained by adjoining the simulated renewal intervals, $s_1 = x_1, \ldots, s_n = x_1 + \ldots + x_n$ (see (2.1.1)). Thus the associated sample function of the associated *counting process* $N(t)$ is also determined (see (2.1.3)): $N(t) = n$ in the interval $[s_n, s_{n+1})$, $n = 0, 1, 2, \ldots$; $s_0 = 0$.   □

The simulation of the counting process in Example (*c*) contains a general principle for the simulation of jump processes. In the simulation we advance from one jump to the next. If $\{X_t, t \in [0, \infty)\}$ is a *jump process* we can associate with it a sequence of random variables $T_0 = 0 < T_1 < T_2 < \ldots$, which represent the randomly distributed *jumps* of the process. Furthermore we may introduce the random variables

$$Y_0 = X_0, \quad Y_n = X_{T_n + 0}, \quad n = 1, 2, \ldots, \tag{1}$$

which denote the value of X_t between the nth and $(n + 1)$th jump. If we succeed in simulating the stochastic process $\{(T_n, Y_n), n = 0, 1, 2, \ldots\}$ with *discrete parameter*, we have also constructed a sample of the process $\{X_t, t \in [0, \infty)\}$, since $X_t = Y_n$ for $T_n \leqslant t < T_{n+1}$. We simulate a jump process by successively constructing the sequence of jumps and by specifying at the same time, when determining each jump, the new value of the jump process in the following interval of constancy.

As an application let us consider *Markov chains with continuous parameter*. Let us assume a *differential matrix* A with the elements q_i and q_{ij} (see Section 3.3), and let us construct sample functions of the associated Markov chain with continuous parameter. At the end of Section 3.3 we have shown that the Markov chain can be identified with a *system of exponentially distributed durations* which has the *same* transition probability functions as the Markov chain, provided that the backward equations have a unique 、 solution. The system with exponentially distributed durations can, however, be understood as a *jump process*. It follows from the definition of the system in Section 3.3 that $\{Y_n, n = 0, 1, 2, \ldots\}$ is a *discrete Markov chain* with the transition probabilities $p_{ij} = q_{ij}/q_i$ for $i \neq j$ and $p_{ii} = 0$. The Markov chain can therefore be simulated by the above scheme as follows: Y_0 is simulated according to the *initial distribution* $p_i, i = 1, 2, \ldots$. If i is a sample of Y_0, then we simulate the duration of state i by the *exponential distribution* with parameter q_i and we have thus obtained a sample of the first jump time T_1. Y_1 is simulated by the transition probabilities $p_{ij}, j = 1, 2, \ldots$, at the same time. If we obtain j as sample of Y_1, then the time until the next jump is simulated according to the exponential distribution with parameter q_j and we obtain a sample for T_2 etc..

The identification of the continuous Markov chain with the jump process with exponentially distributed durations is based on the uniqueness of the solution of the backward equations. Let us therefore investigate this question of uniqueness a little bit further. Let $p_{ij}^{(n)}(t)$ be the probability of a transition from the initial state i at time 0 to the state j at time t *in at most n jumps*. We have $p_{ij}^{(0)}(t) = \delta_{ij}\, e^{-q_i t}$ where $\delta_{ij} = 0$ for $i \neq j$, $\delta_{ij} = 1$ for $i = j$. If the transition from i to j occurs in at most $n + 1$ jumps, then it either occurs without jump or a first jump occurs to a state k at time $s < t$, and from the state k at time s the transition to j at time t occurs in at most n jumps. Thus

$$p_{ij}^{(n+1)}(t) = p_{ij}^{(0)}(t) + q_i \int_0^t e^{-q_i s} \sum_{k \neq i} \frac{q_{ik}}{q_i} p_{kj}^{(n)}(t-s)\, ds. \tag{2}$$

It follows not only for the probabilistic but also for purely analytic reasons from (2) that $p_{ij}^{(0)}(t) \leqslant p_{ij}^{(1)}(t) \leqslant p_{ij}^{(2)}(t) \leqslant \ldots$, and hence there exist the limits

$$p_{ij}(t) = \lim_{n \to \infty} p_{ij}^{(n)}(t) \geqslant 0. \tag{3}$$

Furthermore we have $p_{i1}^{(0)}(t) + p_{i2}^{(0)}(t) + \ldots = e^{-q_i t} \leqslant 1$. Thus it follows from (2) by induction on n that

$$\sum_j p_{ij}^{(n+1)}(t) \leqslant e^{-q_i t} + \int_0^t q_i e^{-q_i s}\, ds = 1$$

and thus

$$\sum_j p_{ij}(t) \leqslant 1. \tag{4}$$

If we let n tend to ∞ in (2), we can see that $p_{ij}(t)$ satisfies equations (3.3.18) and (3.3.19), i.e. the backward equations. It follows from these equations that every other positive solution $r_{ij}(t)$ of them satisfies $p_{ij}^{(0)}(t) \leqslant r_{ij}(t)$ and it follows from (2) by induction on n that $p_{ij}^{(n)}(t) \leqslant r_{ij}(t)$ for all n and finally that $p_{ij}(t) \leqslant r_{ij}(t)$. Therefore $p_{ij}(t)$ is called the *minimal* solution of the backward equations.

If we have *equality* in (4), then $p_{ij}(t)$ is the *unique* solution of the backward equations which represents a transition probability function; for, if

$r_{ij}(t)$ were another positive solution which satisfies (4), we would have to have

$$1 = \sum_j r_{ij}(t) \geqslant \sum_j p_{ij}(t) = 1,$$

and, since $r_{ij}(t) \geqslant p_{ij}(t)$, it follows that $r_{ij}(t) = p_{ij}(t)$. A solution of the backward equations which satisfies (4) with equality, is called *stochastic*, a solution which satisfies (4) with strict inequality is on the other hand called *substochastic*. Substochastic solutions also have a probabilistic interpretation; see Example (e) below. *If the minimal solution is stochastic, then the backward equations have a unique solution which satisfies the necesssary conditions for a transition probability function.* In this case the identification of the Markov chain with the jump process with exponentially distributed durations is justified.

In Section 3.3 we have already mentioned that in applications we may rely on the fact that the backward equations have a unique solution. The following theorem gives a sufficient condition for this.

Theorem 1
If the q_i are bounded, then the minimal solution is stochastic and thus the unique probabilistic solution of the backward equations.

Proof. According to the assumptions we have $q_i \leqslant q < \infty$ for all i and we show by induction on n that

$$\sum_j p_{ij}^{(n)}(t) \geqslant 1 - (1 - e^{-qt})^n. \tag{5}$$

(5) holds for $n = 0$. If we assume (5) for n and sum (2) over j, it follows that

$$\sum_j p_{ij}^{(n+1)}(t) \geqslant e^{-q_i t} + \int_0^t q_i e^{-q_i s}(1 - (1 - e^{-qt})^n) ds$$

$$\geqslant 1 - (1 - e^{-qt})^{n+1},$$

since the right hand side of (5) is monotonically decreasing in t. Therefore (5) holds for all n and $p_{ij}(t)$ must be stochastic. ∎

Example
(d) *Queueing systems.* The queueing systems of Example (3.3.c) satisfy the conditions of Theorem 1. If there are i units in the system, then we simulate the duration of the state i by the exponential distribution with parameter $q_i = \lambda + i\mu$ (if i is less than the number of servers m), or $q_i = \lambda + m\mu$, respectively (if $i \geqslant m$). At the end of the duration of state i we use canonical random numbers for the probabilities λ/q_i and $\min(i, m)\mu/q_i$ to determine whether the system changes into the state $i + 1$ (arrival of a unit)

or $i - 1$ (departure of a unit), and this procedure is repeated subsequently. With this method we compute a sample of the jump process $\{X_t, t \in [0, \infty)\}$, where X_t represents the number of units in the system at time t. The machine maintenance system, Example $(3.3.d)$, can be simulated analogously. □

There exist cases in which the minimal solution is *substochastic*. According to (3) $p_{ij}(t)$ is the probability for a transition from i to j in *finitely* many jumps and accordingly

$$1 - \sum_j p_{ij}(t)$$

is the probability for infinitely many jumps up to time t. In this case there must be at least one accumulation point of jumps before t, i.e. there must exist a more complex discontinuity of the sample functions than a simple jump. This shows that in this case the Markov chain can no longer be identified with a simple jump process. The minimal solution is not the only possible solution any more, either. We can, of course, always simulate a jump process according to the method presented above but we shall obviously not be able to construct the sample path beyond the accumulation point of jumps.

Example

(*e*) *Diverging birth processes.* If we consider pure birth processes (see Example $(3.3.b)$) whose birth rates λ_i for $i = 0, 1, 2, \ldots$ increase to such an extent that

$$\sum_{i=0}^{\infty} \frac{1}{\lambda_i} < \infty,$$

then this series represents precisely the *expected value* of the sum of all successive durations of the states $i = 0, 1, 2, \ldots$ of the associated systems with exponentially distributed durations. Since this expected value is finite, the jumps in the system with exponentially distributed durations must accumulate with probability 1 at a finite random time t. This system obviously corresponds to the minimal solution which is *substochastic*. □

In practical applications we frequently have to simulate far more complicated processes than just simple Markov chains. Complicated *queueing systems*, for example, have to be examined very often by simulation. Let us therefore consider the simulation of a $GI/G/s$-system (see Section 4.1) as another example. The method of Example (*d*) can only be applied to the special case of an $M/M/s$-system; for the general case new methods have to be developed.

A first possibility for the simulation of a $GI/G/s$-system is obtained by considering the waiting times W_n of the units $n = 1, 2, \ldots$ from their time of arrival until the *end* of their service. We consider the situation of the system immediately after the arrival of the nth unit which is assumed to occur at time T_n.

If we base the order of service on the FIFO discipline (see Section 4.1), then all units arriving later are served later. Conversely, the arriving unit n has to wait until all the earlier units have been served and there is a server free to serve it. Let $W_n^{(i)}$ be the length of time from T_n until the server $i, i = 1,$ $2, \ldots, s$ has served the last of the units which have arrived before the unit n. If at time T_n the server i is idle, then we write $W_n^{(i)} = 0$. The unit has to wait until time

$$\min_{i=1,\ldots,s} (W_n^{(i)}) \tag{6}$$

when its service can start. The index j for which the minimum in (6) is achieved describes the server which will be serving the unit n. The service itself lasts time Y which is distributed according to the service distribution $F_B(y)$. The waiting time of the nth unit is thus

$$W_n = Y + \min_{i=1,\ldots,s} (W_n^{(i)}). \tag{7}$$

The next unit $n + 1$ arrives after a time X which is distributed according to the inter-arrival distribution $F_A(x)$. How long does it take from the arrival of the $(n + 1)$th unit until the s servers have finished serving the units which arrived before (including the unit n)? In other words, how do we determine the $W_{n+1}^{(i)}, i = 1, 2, \ldots, s$? Since time X has elapsed since the arrival of the last unit n, we have, for the server $i \neq j$, $W_{n+1}^{(i)} = W_n^{(i)} - X$ if $W_n^{(i)} \geqslant X$; otherwise $W_{n+1}^{(i)}$ is obviously equal to 0. For the server j, which serves or will be serving the unit n, we obviously have $W_{n+1}^{(j)} = W_n - X$ or 0, depending on whether $W_n \geqslant X$ or $W_n < X$, since this server will be idle at time $T_n + W_n$. Summarizing we thus have

$$W_{n+1}^{(j)} = \max(0, W_n - X),$$

$$W_{n+1}^{(i)} = \max(0, W_n^{(i)} - X), \quad i = 1, 2, \ldots, s, \; i \neq j. \tag{8}$$

The relations (6) to (8) now allow the simulation of the $GI/G/s$-system. The simplest way to begin the simulation is to assume that at the arrival of the first unit (number 0) all servers are idle, i.e. no units have entered the system. Then we write $W_0^{(i)} = 0$ for all $i = 1, \ldots, s$ and let the first unit arrive at time $T_0 = 0$. The further simulation can be done by repetition of the following computational steps.

1. Simulate the inter-arrival time X by $F_A(x)$ and the service time Y by $F_B(y)$ with the methods of Section 6.2.
2. Determine the waiting time W_n of the nth unit by (7).
3. Determine the arrival time of the next unit $T_{n+1} = T_n + X$ and the new values of $W_{n+1}^{(i)}$ by (8) from $W_n^{(i)}$ and W_n, respectively.

With this computational method we obtain samples of the arrival times t_n, the service times y_n and the waiting times w_n of the units $n = 1, 2, \ldots$. This computation can, of course, not be made for infinitely many units but has to terminate after a certain number N of units. From the above samples we can, however, obtain a number of further samples of quantities which are of interest in connection with the $GI/G/s$-system.

Thus the waiting time of the nth unit from the time of arrival up to the beginning of the service is equal to $w_n - y_n$. Furthermore we can construct the sample function of the process X_t of the number of units which are in the system at time t. The sample functions of X_t are *step functions* which at the arrival times $t_n, n = 1, 2, \ldots$, jump up by one. At the end of a service, i.e. at the times $t_n + w_n$, the step functions jump down by one. The sample function of X_t is uniquely determined by the samples t_n, w_n of T_n and W_n.

When applying this method the simulation of the $GI/G/s$-system is basically reduced to the simulation of the processes $\{W_n^{(1)}, \ldots, W_n^{(s)}, n = 1, 2, \ldots\}$ with *discrete* parameter. There are other possibilities for the simulation. Greatly generalized concepts have been developed for the simulation of complex stochastic processes. Based on these, special computer programming languages, so-called *simulation languages*, have even been developed. These contain pre-programmed routines for the generation of random numbers, for the simulation of random variables with important special distributions as well as for the representation and statistical processing of the simulated samples.

6.4. Monte-Carlo methods

If one succeeds in representing the solution of a mathematical problem as a parameter of a probability experiment or of a stochastic process, then this solution can be found by *simulating* the experiment or process in question and by estimating the unknown parameter by *statistical analysis* of the simulation results. This method of solving a problem is called a *Monte-Carlo method*. In most applications the problem to be solved is in itself already of a probabilistic nature, so that the solution is immediately given as a parameter of a stochastic process. However, the Monte-Carlo method can sometimes be applied to problems which initially have nothing to do with probabilistic questions. It is furthermore possible that the sought solution can be connected with several different probabilistic experiments. In this case we can choose the most convenient experiment for the Monte-Carlo method.

In the simplest case the desired solution is a *real number* θ which can be represented as the *expected value* of a *random variable* Y with the distribution function $F(y)$:

$$\theta = \int_{-\infty}^{\infty} y \, dF(y). \tag{1}$$

Every random variable Y whose expected value is equal to θ (or at least sufficiently close to θ) is called a *primary solution* for θ. The distribution function $F(y)$ of a primary solution does not have to be known explicitly. It is sufficient that we can represent or write Y as a function f of the random variables $X_1, X_2, \ldots$ of the canonical experiment; $Y = f(X_1, X_2, \ldots)$. By substituting *canonical random numbers* $x_1, x_2, \ldots$ for $X_1, X_2, \ldots$ we obtain *samples* $y = f(x_1, x_2, \ldots)$ of Y. In general we only need a *finite* number of random numbers $x_1, x_2, \ldots, x_h$ for the computation of y where h, however, can depend on the sample — we only have to think of the rejection method (Section 6.2). To simplify matters let us assume that there exists a finite upper bound N for h so that we can write $Y = f(X_1, X_2, \ldots, X_N)$. If $x = (x_1, x_2, \ldots, x_N)$ and $dx = dx_1 dx_2 \ldots dx_N$, then

$$\theta = \int f(x) dx, \tag{2}$$

where the integration domain is the N-dimensional *unit cube* $0 \leqslant x_i \leqslant 1, i = 1, \ldots, N$.

If we consider a sequence of *independent* random variables $Y_1, Y_2, \ldots,$ $Y_n, n = 1, 2, \ldots,$ which are all *identically* distributed primary solutions for θ, the *arithmetic mean*

$$\bar{Y}_n = \frac{1}{n} \sum_{i=1}^{n} Y_i \tag{3}$$

is called a *secondary solution* for θ. By taking independent groups x_i of N canonical random numbers $x_1, \ldots, x_N$ each, we obtain samples $y_i = f(x_i)$, $i = 1, \ldots, n$, for the variables $Y_i, i = 1, \ldots, n$. The *sample average*

$$\bar{y}_n = \frac{1}{n} \sum_{i=1}^{n} y_i \tag{4}$$

is a sample of $\bar{Y}_n$ and is called the *estimated value* for θ.

We must, of course, not expect $\bar{y}_n$ to be exactly equal to θ, and it is equally impossible to make an absolute statement about the deviation of $\bar{y}_n$ from θ. We can only make *probabilistic* statements about the probable mistakes $\bar{y}_n - \theta$. For this let us make the most important assumption that the primary solution Y has a *finite variance* σ^2. Then the secondary solution $\bar{Y}_n$ has the variance σ^2/n (see (1.4.17)). In any case $\bar{Y}_n$ has the expected value θ (see (1.4.16)),

$$E[\bar{Y}_n] = \theta, \qquad \text{Var}[\bar{Y}_n] = \sigma^2/n. \tag{5}$$

According to the *central limit theorem* (1.4.22) we have, for sufficiently large n, $\sqrt{n}\,(\bar{Y}_n - \theta)/\sigma$ approximately *normally distributed* with the expected value 0 and variance 1. We thus have

$$P\left[|\bar{Y}_n - \theta| \leqslant \frac{k\sigma}{\sqrt{n}}\right] \approx \frac{1}{\sqrt{(2\pi)}} \int_{-k}^{k} e^{-z^2/2} dz = \Phi(k) - \Phi(-k). \tag{6}$$

This means, for example, that we do not make a mistake in $(\Phi(k) - \Phi(-k)) \times 100\%$ of the cases, if we maintain that the interval $\bar{y}_n \pm k\sigma/\sqrt{n}$ contains the unknown parameter θ. We have, for example, $\Phi(1) - \Phi(-1) = 0.6826$; $\Phi(2) - \Phi(-2) = 0.9544$; $\Phi(3) - \Phi(-3) = 0.9974$, as can be found in tables of the normal distribution. It is quite remarkable that the *inaccuracy of the estimate* in the stated sense *is proportional to the variance of the primary solution and inversely proportional to the root of the sample size n*. The accuracy thus increases only slowly with the number of samples used in the estimate. However, the estimate converges with certainty, for, according to the *law of large numbers*, (1.4.20), we have

$$P[\lim_{n \to \infty} \bar{Y}_n = \theta] = 1. \tag{7}$$

Example
(*a*) *Computation of integrals.* It emerges from (1) and (2) that the Monte-Carlo method can be used for the computation of integrals. In fact, the Monte-Carlo method can be used for the computation of very general integrals. Let us explain this by the simple example of a one-dimensional Riemann integral

$$\theta = \int_{-\infty}^{\infty} g(x)\,dx.$$

Let $f(x)$ be a *density function* with $f(x) \neq 0$, if $g(x) \neq 0$. If we put $h(x) = g(x)/f(x)$, if $f(x) \neq 0$ and $h(x) = 0$, if $f(x) = 0$, we have

$$\theta = \int_{-\infty}^{\infty} h(x)f(x)\,dx.$$

This, however, is the expected value of a random variable $Y = h(Z)$ where Z is a random variable with the density $f(x)$. $Y = h(Z)$ is a *primary solution* for θ. We simulate Y by constructing samples z of Z and computing $y = h(z)$. Note that one obtains *different* primary solutions, depending on the choice of $f(x)$. We are interested in choosing $f(x)$ in such a way that the simulation is as *simple* as possible but also in such a way that the variance of the primary solution is *as small as possible* in order to obtain good accuracy of the Monte-Carlo method. Such Monte-Carlo methods for the computation of integrals become particularly useful in comparison with other numerical methods, for multi-dimensional complicated integrals. The computation of *sums* and *series*, by the way, can also be done quite analogously with a discrete distribution on the summation index instead of the density function $f(x)$. □

From what we have said so far, we can see that, apart from the construction of the required samples, the Monte-Carlo method becomes to a large extent a problem of *mathematical statistics*. In the following we return to the problem of increasing the accuracy of the Monte-Carlo method by constructing suitable primary solutions.

On the basis of the above considerations, it is important, when trying to obtain high accuracy for the basic pattern of the Monte-Carlo method, to find a primary solution with a small variance. Attempts to this effect are called *variance reducing* methods. There exist a number of such methods and the most widely known can be divided into two groups.

The first group consists of the *correlation methods*, which in turn can be divided into *comparison methods* and *compensation methods*. Let Y be a primary solution for the unknown parameter θ and let Z_1, Z_2 be primary solutions for a second parameter τ. Then $W = Z_1 + (Y - Z_2)$ is also a primary solution for θ. Now let $X = (X_1, X_2, \ldots, X_N)$ be a vector of independent, uniformly distributed in $[0, 1]$, random variables where N is sufficiently large such that Y, Z_1, Z_2 can be represented in the form $Y = f(X), Z_i = g_i(X)$. Let X' and X'' be *independent* random vectors and let us define W more accurately by

$$W = g_1(X') + (f(X'') - g_2(X''));$$ (8)

$Y = f(X'')$ and $Z_2 = g_2(X'')$ will generally have a non-zero *covariance*

$$\mathrm{Cov}[Y, Z_2] = \int (f(x) - \theta)(g_2(x) - \tau)\mathrm{d}x.$$ (9)

Z_1 and $(Y - Z_2)$ are, however, mutually *independent*. Therefore the variance of the primary solution W (see (1.4.2) and (1.4.3)) becomes

$$\sigma_W^2 = \sigma_{Z_1}^2 + \sigma_Y^2 + \sigma_{Z_2}^2 - 2\,\mathrm{Cov}[Y, Z_2].$$ (10)

It follows that the variance of W can be kept small in comparison with that of Y if we can introduce a large *positive covariance* between Y and Z_2 and if, in addition, Z_1 has a small variance. We obtain a large covariance between Y and Z_2 by choosing for g_2 a function which is as similar as possible to f, $g_2 \approx f$. Then, on the basis of (9), $\sigma_{Z_2}^2 \approx \mathrm{Cov}[Y, Z_2] \approx \sigma_Y^2$ and the term $\sigma_Y^2 + \sigma_{Z_2}^2 - 2\,\mathrm{Cov}[Y, Z_2]$ is thus kept small.

This method becomes particularly interesting if the parameter τ is known. We then put $Z_1 = \tau$ and obtain $\sigma_{Z_1}^2 = 0$. Then $W = \tau + (Y - Z_2)$ and $\sigma_W^2 = \sigma_Y^2 + \sigma_{Z_2}^2 - 2\,\mathrm{Cov}[Y, Z_2]$.

Thus the method assumes two similar primary solutions ($f \approx g_2$) for two parameters θ and τ, as well as a second good primary solution for τ. By simulating the similar primary solutions for θ and τ with a *large positive covariance* (equal random numbers, see (8)), we can reduce the variance of the original primary solution for θ. This method is called the *comparison method*.

To describe the *compensation method* consider m primary solutions $Y^{(i)}$, $i = 1, \ldots, m$ for θ; then

$$Z = \sum_{i=1}^{m} \lambda_i Y^{(i)}, \quad \sum_{i=1}^{m} \lambda_i = 1 \tag{11}$$

is also a primary solution for θ. Let it again be allowed that the $Y^{(i)}$ are *dependent*: this dependence can have a variance reducing effect. Let $\sigma_{ij} = \sigma_{ji} = \text{Cov}[Y^{(i)}, Y^{(j)}]$ and $\sigma_{ii} = \text{Var}[Y^{(i)}]$, $i = 1, \ldots, m$. We have (see (1.4.2))

$$\sigma_Z^2 = \text{Var}[Z] = \sum_{i=1}^{m} \sum_{j=1}^{m} \sigma_{ij} \lambda_i \lambda_j. \tag{12}$$

Since $\sigma_Z^2 \geqslant 0$ for all λ, σ_Z^2, as a function of λ, must have a minimum under the side condition $\lambda_1 + \ldots + \lambda_m = 1$. If we apply the well-known method of Lagrange multipliers, we find as a condition for this minimum

$$\sum_{j=1}^{m} \sigma_{ij} \lambda_j = c, \quad i = 1, \ldots, m,$$

$$\sum_{j=1}^{m} \lambda_j = 1, \tag{13}$$

where c is an (unknown) constant. If we choose Z as primary solution for θ, then we speak of a *compensation method*.

Usually we start with a given primary solution Y for θ in order to construct further primary solutions $Y^{(1)}, Y^{(2)}, \ldots$. Let $Y = f(X), X = (X_1, X_2, \ldots, X_N)$. Then we look for maps T of the N-dimensional space which carry the N-dimensional unit cube into itself and preserve the uniform distribution of the unit cube, i.e. for which both X and $\hat{X} = T(X)$ are uniformly distributed in the N-dimensional unit cube. If T is such a map and $Y = f(X)$ is a primary solution for θ, then $Y' = f(T(X))$ is also a primary solution for θ. The simplest examples of such maps are given by

(i) $\hat{X}_i = 1 - X_i \quad i = 1, \ldots, N,$

(ii) $\hat{X}_i = \begin{cases} X_i + k/h, & \text{if } X_i + k/h \leqslant 1 \\ X_i + k/h - 1 & \text{if } X_i + k/h > 1, \\ \quad i = 1, \ldots, N; \; k, h \text{ integral}, k < h. \end{cases} \tag{14}$

If $Y = f(X)$ is a primary solution for θ, we may put in case (i) $Y^{(1)} = Y$ and $Y^{(2)} = f(T(X))$. Obviously $\sigma_{11} = \sigma_{22} = \sigma^2$ and it follows from (13) that $\lambda_1 = \lambda_2 = \frac{1}{2}$. We thus choose

$$Z = \tfrac{1}{2}(Y^{(1)} + Y^{(2)}) \tag{15}$$

and obtain (see (1.4.2))

$$\sigma_Z^2 = \tfrac{1}{2}(\sigma^2 + \sigma_{12}). \tag{16}$$

Since $\sigma_{12} \leqslant \sigma^2$ (see (1.3.35)), the variance of the new primary solution Z is in any case less than or equal to the variance σ^2 of the original primary solution. This in itself is, however, not too significant since the complexity of the simulation of (15) corresponds approximately to the one for the secondary solution (3) with $n = 2$. We have thus made genuine progress only if $\frac{1}{2}(\sigma^2 + \sigma_{12}) < \frac{1}{2}\sigma^2$, i.e. if $\sigma_{12} < 0$. Whether this is the case depends on the function f. Since we replace the small (large) random numbers by large (small) random numbers with the transformation (14) (i) we may often be justified in hoping that the random numbers will behave in an opposite manner and thus lead to a negative covariance.

Let h be fixed in case (ii) and let the map (14) (ii) with $k = 0, 1, 2, \ldots,$ $h-1$ be denoted by $T^{(k+1)}$. Furthermore, let $Y^{(i)} = f(T^{(i)}(X)), i = 1, \ldots, h$. We have $\sigma_{ii} = \sigma^2$ for $i = 1, \ldots, h$ and $\sigma_{ij} = \sigma_{|i-j|}$, i.e. the covariance of $Y^{(i)}$ and $Y^{(j)}$ depends only on $|i-j|$. Furthermore we also have $\sigma_{|i-j|} = \sigma_{h-|i-j|}$ by (14) (ii); the reader is advised to convince himself by considering the cases $h = 3, 4, \ldots$. Thus it follows from (13), as well as for reasons of symmetry, that $\lambda_i = 1/h, i = 1, \ldots, h$, and

$$Z = \sum_{i=1}^{h} \frac{f(T^{(i)}(X))}{h}, \tag{17}$$

$$\sigma_Z^2 = \frac{\sigma^2 + 2 \sum_{k=1}^{h} \frac{h-k}{h} \sigma_k}{h}. \tag{18}$$

For the same reason as in case (i) σ_Z^2 ought not to be compared with the variance σ^2 of the primary solution Y but with σ^2/h. We thus prefer Z to Y if $2((h-1)\sigma_1 + (h-2)\sigma_2 + \ldots + \sigma_{h-1})/h < 0$. (i) and (ii) represent special compensation methods which are also called methods of *antithetical variables*.

All these variances and covariances which are necessary for the decision whether a variance reduction can be obtained by these methods are never really known in advance. Unless one wants to estimate the quantities in question in advance by additional considerations, one has to rely on intuitive judgement and decide whether it is worthwhile or not to use one of the methods presented here.

Examples

(b) *Antithetical random numbers.* The sequence of random numbers which is generated by the *multiplicative congruence method*, is uniquely determined by the *seed* x_0' (see Section 6.1) for a given factor a and a given module m. If we consider in addition the second sequence of random numbers arising from the seed $x_0'' = m - x_0'$, then we have for the next number of this sequence

$$x_1'' \equiv ax_0'' \equiv a(m - x_0') \equiv m - ax_0' \equiv m - x_1' \pmod{m}.$$

For the next two and thus for all the following numbers of the two sequences we have $x_i'' = m - x_i'$. Thus we can see, after dividing by m that we obtain *antithetical* sequences of random numbers $\{x_1, x_2, \dots\}$ and $\{1 - x_1, 1 - x_2, \dots\}$ in the sense of (14) (i), if we start with the antithetical seeds x_0' and $m - x_0'$. □

(c) *Determining the differences of parameters.* Let θ_1 and θ_2 be two un-known parameters and let the task be the determination of the *difference* $\theta_1 - \theta_2$ of the two parameters. Such problems arise in particular in the optimization of systems when the problem is how to compare two different systems with respect to a certain criterion.

If $Y_1 = f_1(X_1, X_2, \dots)$ and $Y_2 = f_2(X_1, X_2, \dots)$ are *primary solutions* for θ_1 and θ_2, then $Y_1 - Y_2$ is a *primary solution* for $\theta_1 - \theta_2$. If Y_1 and Y_2 have the variances σ_1^2 and σ_2^2, then $Y_1 - Y_2$ has the variance $\sigma_1^2 + \sigma_2^2$ if Y_1 and Y_2 are independent. In simulation this independence is obtained by simu-lating Y_1 and Y_2 with non-overlapping parts of the random number sequence.

A variant of the *comparison method* would consist in simulating Y_1 and Y_2 with the *same* random numbers. In this case it is quite often possible that we obtain considerable *positive* covariance $\mathrm{Cov}[Y_1, Y_2] > 0$; this can, how-ever, generally not be guaranteed. The variance of $Y_1 - Y_2$ becomes $\sigma_1^2 + \sigma_2^2 - 2\,\mathrm{Cov}[Y_1, Y_2]$, and we obtain a *variance reduction* for the estimate of the difference $\theta_1 - \theta_2$ if $\mathrm{Cov}[Y_1, Y_2] > 0$. However, this method yields no variance reduction for the estimates of θ_1 and θ_2, respectively.

Of course, we can also obtain a variance reduction for $\theta_1 - \theta_2$ if we suc-ceed in reducing the variances σ_1^2 and σ_2^2 for the estimates of θ_1 and θ_2. One should be careful when combining variance reducing methods for θ_1 and θ_2 with the above comparison method since it could happen that, through com-bination, undesirable correlations could be introduced which could destroy the variance reduction for $\theta_1 - \theta_2$. □

The second group of the variance reducing methods can be subsumed under the *method of the essential sample* (importance sampling).

Let, as at the beginning of this section, X be an N-dimensional vector and $Y = f(X)$ a *primary solution* for the unknown parameter θ. Let $h(x)$ be a density function such that $h(x) \neq 0$ if $f(x) \neq 0$ for all x of the N-dimensional unit cube $0 \leqslant x_i \leqslant 1$. Let the ratio $f(x)/h(x)$ be equal to zero if $f(x) = 0$. It follows from (2) that

$$\int \frac{f(x)}{h(x)} h(x)\,dx = \theta \tag{19}$$

(all integrals are taken over the N-dimensional unit cube). Hence, $Z = f(X)/h(X)$ is also a *primary solution* for θ where the vector X is now distributed with density $h(x)$. The variance of Z is

$$\sigma_Z^2 = \int \frac{f^2(x)}{h^2(x)} h(x)\,\mathrm{d}x - \theta^2. \tag{20}$$

If $f(x) \geq 0$, then

$$h(x) = \frac{f(x)}{\int f(x)\,\mathrm{d}x} = \frac{f(x)}{\theta} \tag{21}$$

is a density function which satisfies the necessary conditions, and with this density the variance (20) *vanishes.* In this case the entire variance would vanish. In practice this can, of course, not be done. In order to apply the density (21) the unknown parameter θ would already have to be known. This consideration shows, however, that a considerable variance reduction can be obtained if one chooses a density $h(x)$ which is at least *approximately proportional* to $f(x)$.

When simulating by such a density function $h(x)$ we are likely to obtain sample values x in regions in which $f(x)$ is *large.* This means that we concentrate the samples in regions in which we may expect significant contributions to the integral (2) for θ. This is the reason for calling this a method of essential samples or importance sampling.

The successful application of these variance reducing methods in Monte-Carlo studies, as one would come across them in practical applications, is anything but simple. For probabilistic problems a primary solution Y is usually naturally determined. It is also relatively simple to write a computer program which can generate samples of the primary solution with the help of canonical random numbers. Hence, function $Y = f(X)$ is *implicitly* defined. $f(x)$ is frequently not a very smooth function of x, and generally quite complex. Therefore the above attempts to reduce the variances can only be successful if combined with a clever exploitation of the properties of the individual special case. In Examples (b), (c) and (d) of the following section (6.5) we shall show how this is done. The methods presented here are useful models; their aim is to bring to the reader's attention that the 'natural' primary solution is not necessarily the only and not necessarily the best solution for his problem.

6.5. Regenerative simulation

The basic pattern of the Monte-Carlo method presented in Section 6.4 is not the only possibility of determining unknown parameters by simulation. There exist problems for which different methods are more advantageous. As an example we consider a *discrete Markov chain* $\{X_n, n = 0, 1, 2, \ldots\}$ with *an aperiodic positively recurrent class.* Let $I = \{1, 2, \ldots\}$ be the state space of the chain and let $r : I \to \mathbb{R}$ be a map of the state space into the set of real numbers. Under these conditions the chain has a uniquely determined *stationary distribution* $p_i, i = 1, 2, \ldots$ (see Theorem 3.2.9). $r(i)$ can be considered as pay-off in state i. Let us now determine the *profit rate* (see Section 5.3)

$$g = \sum_i r(i) p_i .\tag{1}$$

Suppose the chain has the transition probabilities p_{ij} and the n-step transition probabilities $p_{ij}^{(n)}$. According to Theorem 3.2.9 the $p_{ij}^{(n)}$ tend to p_j as $n \to \infty$, for every initial state i. If the $r(i)$ are *bounded* then

$$g_n = \sum_j r(j) p_{ij}^{(n)}\tag{2}$$

also converges to g as $n \to \infty$. $p_{ij}^{(n)}$ is the distribution of X_n, given that $X_0 = i$. By Section 6.4, $r(X_n)$ is, for sufficiently large n, a primary solution for the unknown parameter g. The *arithmetic mean* of the g_n over $h = 0, \ldots, n$ also converges to g as $n \to \infty$; thus

$$\frac{1}{n+1} \sum_{h=0}^{n} r(X_h)\tag{3}$$

is also a primary solution of g.

According to the Monte-Carlo method the Markov chain would have to be simulated repeatedly up to a sufficiently large n by the well-known method (Section 6.3): every time we have to construct the average (3) and then form the arithmetic mean of these results of the individual simulation.

We shall now develop another method for which it is not necessary to repeat the simulation of the Markov chain from a given initial state. Let i be an arbitrary recurrent state and $X_0 = i$. We consider the *renewals* of the chain in state i and define the renewal intervals $Y_h, h = 1, 2, \ldots$, as the number of transitions from the moment in which the chain has reached state i for the $(h-1)$th time, up to the hth time when it reaches state i again. The Y_h are mutually independent, have the same distribution $P[Y_h = k] = f_i^{(k)}$ and define a *discrete recurrent renewal process* (see Section 3.2).

We have $E[Y_h] = \mu_i = 1/p_i$ (see Theorem 3.2.9). Furthermore let

$$Z_h = \sum_{n=Y_1+\ldots+Y_{h-1}}^{Y_1+\ldots+Y_h-1} r(X_n), \quad h = 1, 2, \ldots,\tag{4}$$

be the sum of the pay-offs from the $(h-1)$th up to the hth renewal in state i.

The sequence $\{(Z_h, Y_h), h = 1, 2, \ldots\}$ defines a *cumulative process* (see Section 2.3, in particular (2.3.23))

$$V(n) = \sum_{h=1}^{N(n)+1} Z_h,\tag{5}$$

where $N(n)$ is the counting process of the renewals of the Markov chain in state i. It follows from (4) and (5) that

$$E[V(n)] = E\left[\sum_{j=0}^{n} r(X_j)\right] + E\left[\sum_{j=n+1}^{m} r(X_j)\right], \tag{6}$$

where the second sum goes up to $m = Y_1 + \ldots + Y_{N(n)+1} - 1$. The second term on the right hand side of (6) is less than $E[Z_h]$, and this expected value is finite if the $r(i)$ are bounded. The first term on the right hand side of (6) is equal to $g_0 + g_1 + \ldots + g_n$. If we divide (6) by n, it follows from Theorem 2.3.5 that

$$\frac{E[Z_1]}{E[Y_1]} = \lim_{n \to \infty} \frac{V(n)}{n} = \lim_{n \to \infty} \frac{1}{n} \sum_{j=0}^{n} g_j = g. \tag{7}$$

Since $Z_1, Z_2 \ldots$ are *independent* random variables with the same distribution and in particular with the same expected value $E[Z_h] = E[Z_1], h = 2, 3, \ldots$, we have the *strong law of large numbers* (see (1.4.20))

$$\lim_{m \to \infty} \frac{1}{m} \sum_{h=1}^{m} Z_h = E[Z_1] \tag{8}$$

with probability 1. Similarly we also have

$$\lim_{m \to \infty} \frac{1}{m} \sum_{h=1}^{m} Y_h = E[Y_1] \tag{9}$$

with probability 1. Hence we obtain from (7) with probability 1 that

$$\lim_{m \to \infty} \frac{\sum_{h=1}^{m} Z_h}{\sum_{h=1}^{m} Y_h} = g. \tag{10}$$

The result (10) allows the following simulation method for the estimation of the profit rate g: choose an arbitrary recurrent state i and simulate the Markov chain from $X_0 = i$ until a sufficiently large number m of renewals in state i have occurred. Then, according to (10), the sum of the pay-offs r over all simulated transitions, divided by the number of the simulated transitions, is an approximation to g.

Example

(a) *Monte-Carlo determination of the stationary distribution.* If we put $r(i) = 1$ and $r(j) = 0$ for $j \neq i$, then $g = p_i$ and the above method is used to estimate the probability p_i of the state i in the stationary distribution. ☐

This method can be applied to more general systems than the discrete Markov chain. The only important condition is that there exist (random) times $T_0 = 0 < T_1 < T_2 < \ldots$ which form a *recurrent renewal process* such that the developments of the system are mutually *independent* during the renewal intervals $[T_{h-1}, T_h)$ and are governed by the same probability laws. At every $T_h, h = 0, 1, 2, \ldots$, the same probabilistic development starts again. The system is self-regenerative (see also Chapter 7). If we make use of this property, we speak of a *regenerative simulation*. There exist, as in the example of Markov chains, cumulative processes to which Theorem 2.3.5 can be applied. The desired quantity is obtained as the ratio of two expected values − as in (7) − which can be estimated by the law of large numbers.

We come across such regenerative systems quite frequently: consider the example of a $GI/G/s$-system. As shown in Section 4.1, the times when the arriving units find all servers idle are such periods of regeneration for the $GI/G/s$-system. Consequently, Theorem 4.1.1 holds. This theorem permits the determination of the average length of the queue

$$\lim_{t \to \infty} \frac{1}{t} E\left[\int_0^t X_t \, dt\right] = L \tag{11}$$

and the *average waiting time* of the units

$$\lim_{n \to \infty} \frac{1}{n} \sum_{i=0}^{n} E[W_i] = W. \tag{12}$$

The intervals between two successive periods of regeneration have been called *cycles* in Section 4.1 and we have introduced the random variables Z_i (duration of the ith cycle), Y_i (the number of units served during the ith cycle) and V_i (sum of the waiting time of the served units in the ith cycle). We have by Theorem 4.1.1

$$L = E[V_1]/E[Z_1], \qquad W = E[V_1]/E[Y_1]. \tag{13}$$

Since the $V_i, i = 1, 2, \ldots$, as well as the Z_i and $Y_i, i = 1, 2, \ldots$, are each mutually independent and have the same expected values $E[V_i] = E[V_1]$, $E[Z_i] = E[Z_1], E[Y_i] = E[Y_1], i = 1, 2, \ldots$, it follows from the *strong law of large numbers* that

$$\lim_{m \to \infty} \frac{1}{m} \sum_{i=1}^{m} V_i = E[V_1],$$

$$\lim_{m \to \infty} \frac{1}{m} \sum_{i=1}^{m} Z_i = E[Z_1],$$

$$\lim_{m \to \infty} \frac{1}{m} \sum_{i=1}^{m} Y_i = E[Y_1] \tag{14}$$

with probability 1 and thus also with probability 1 that

$$\lim_{m \to \infty} \frac{\sum_{i=1}^{m} V_i}{\sum_{i=1}^{m} Z_i} = L, \quad \lim_{m \to \infty} \frac{\sum_{i=1}^{m} V_i}{\sum_{i=1}^{m} Y_i} = W. \tag{15}$$

The $GI/G/s$-system has to be simulated over a sufficiently large number m of cycles. The ratios in (15) can then be regarded as estimates for L and W. Similarly we can obtain estimates for further quantities which might be of interest for the $GI/G/s$-system.

Examples

(b) *Variance reduction for regenerative simulation.* The variance for an estimate like (10) cannot be expressed in terms of variances of the associated random variables Z_h and Y_h. Besides, the expected values of the estimates (10) for $m = 1, 2, \ldots$ are in general not equal to the unknown parameter g. This makes the investigation of variance reducing methods more difficult. Nevertheless the ideas of Section 6.4 can be carried over to the case of regenerative simulation.

We can, instead of investigating the actual variance $E[((\Sigma Z_h/\Sigma Y_h) - E[\Sigma Z_h/\Sigma Y_h])^2]$ of the estimate (10), just as well examine the expected squared deviation from the unknown parameter g, $E[((\Sigma Z_h/\Sigma Y_h) - g)^2]$.

If we have a second regenerative simulation with the estimated values

$$\lim_{m \to \infty} \frac{\sum_{h=1}^{m} Z'_h}{\sum_{h=1}^{m} Y'_h} = g',$$

and if g' is *known*, then

$$\lim_{m \to \infty} \left\{ \frac{\sum_{h=1}^{m} Z_h}{\sum_{h=1}^{m} Y_h} - \frac{\sum_{h=1}^{m} Z'_h}{\sum_{h=1}^{m} Y'_h} + g' \right\} = g$$

are also estimates for g (see (6.4.8), where $g_1(X')$ has been replaced by the parameter τ which is assumed to be known). If we succeed in correlating the two simulations by using the same random numbers such that we have considerable *positive* covariance between the two ratios $(\Sigma Z_h/\Sigma Y_h)$ and $(\Sigma Z'_h/\Sigma Y'_h)$, then we may expect, as for the *comparison method* of Section 6.4, an improvement in accuracy for the estimate of g. It is sometimes possible to construct the two simulations in such a way that $Y_h = Y'_h$. If, in addition, we can obtain $Z_h \approx Z'_h$ with large probability one has surely achieved an improvement. An application is given in the following Example (c).

We can furthermore construct groups of k successive regenerating cycles and attempt to correlate the random variables $Z_{hk+1}, \ldots, Z_{(h+1)k}$ and $Y_{hk+1}, \ldots, Y_{(h+1)k}$ within these groups as for the *compensation method* (see Section 6.4), such that the *group sums*

$$\sum_{j=hk+1}^{(h+1)k} Z_j, \quad \sum_{j=hk+1}^{(h+1)k} Y_j, \quad h = 0, 1, 2, \ldots,$$

obtain smaller variances than if the random variables were independent. Since the composition of k regenerative cycles again forms a regenerative cycle and these composed cycles are mutually *independent*, as is assumed for the regenerative simulation, we have also

$$\lim_{m \to \infty} \frac{\sum_{h=0}^{m} \sum_{j=hk+1}^{(h+1)k} Z_j}{\sum_{h=0}^{m} \sum_{j=hk+1}^{(h+1)k} Y_j} = g$$

and we may expect that the cancellation within the groups improves the accuracy of the estimate g. We describe an application in Example (d). $\qquad\square$

(c) *Variance reduction for Markov chains.* Let us, as at the beginning of the section, determine the profit rate g of the pay-offs $r(i)$. Suppose, for reasons of simplicity, that the state space is *finite*, so that we can apply the results of Section 5.3. If we have, from a preliminary investigation, a first rough estimate g' of the profit rate and a rough estimate v_i' of the *relative values* (see Section 5.3) for the pay-offs $r(i)$, then we can use this knowledge in a second step. Since these estimates will contain *errors*, equation (5.3.10) will not be satisfied exactly. We have to introduce a correction term $k(i)$ in the ith equation of (5.3.10) so that we have

$$g' + v_i' = r(i) + k(i) + \sum_{j=1}^{N} p_{ij} v_j'.$$

However, it follows from this that g' is the exact profit rate for the pay-offs $r(i) + k(i)$ (see (5.3.11)). With the *comparison method* (Example (b)) we can thus simulate not only the original Markov chain with the pay-offs $r(i)$ but also the *same* Markov chain with the pay-offs $r(i) + k(i)$. The correlation between the two simulations is established by simulating the chain itself only *once*. In (10) we then have $Y_h = Y_h'$ and the Z_h, Z_h' are obtained from the different weighting of the states i with $r(i)$ and $r(i) + k(i)$, respectively. We then obtain

$$Z_h - Z_h' = \sum_{n=Y_1+\ldots+Y_{h-1}}^{Y_1+\ldots+Y_{h-1}} k(X_n) = Z_h''$$

(see (4)). What this comparison method amounts to is that we simulate the *original* Markov chain with the *new* pay-offs $k(i)$ and we estimate the profit rate g by

$$\lim_{m \to \infty} \frac{\sum_{h=1}^{m} Z_h''}{\sum_{h=1}^{m} Y_h} + g' = g.$$

If the errors $k(i)$ are small in comparison with the differences of the $r(i)$, we may expect an improvement in accuracy. $\qquad\square$

(d) *Variance reduction for queueing systems.* It may be advantageous for the simulation of queueing systems (e.g. the $GI/G/s$-system) to organize the computation in such a way that for the generation of inter-arrival times and the sequence of service times of each server, *separate* subsequences of canonical random numbers should be applied. This can be done quite easily by using *different random number seeds* $x_0^{(1)}, x_0^{(2)}, \ldots, x_0^{(s+1)}$ for the sequences of the inter-arrival times and of the service times of the servers 1 to s.

Let us denote the $s + 1$ different random number sequences which arise from the seeds $x_0^{(1)}, \ldots, x_0^{(s+1)}$ by $F_1, \ldots, F_{s+1}$. For each of these sequences F_i there exists, according to (6.4.14) (i), an *antithetical* sequence A_i which arises from the seed $m - x_0^{(i)}$ (Example (6.4.b)). Regeneration cycles can be grouped together with all possible combinations of the antithetical random number sequence in a *compensation procedure* (see Example (b)). A first cycle which is simulated by the sequences $F_1, F_2, \ldots, F_{s+1}$, can be followed by a second cycle with $A_1, F_2, \ldots, F_{s+1}$. In doing so, long (short) intervals change into short (long) intervals, while the service times remain the same. Furthermore, the cycles can be followed by antithetical service time sequences, for example, $F_1, A_2, F_3, \ldots, F_{s+1}; F_1, F_2, A_3, \ldots, F_{s+1}; F_1, F_2, \ldots, A_{s+1}; F_1, A_2, A_3, \ldots, F_{s+1}; \ldots; F_1, A_2, A_3, \ldots, A_{s+1}$, etc.. In all there exist 2^{s+1} different combinations of antithetical sequences. For increasing s this number becomes very quickly too big to construct groups with all possible combinations. One has to restrict oneself to a small number of suitable combinations. The choice of such combination is a question of *planning of experiments*.

Frequently two different queueing systems have to be compared; for example, the average waiting times for the $GI/G/(s-1)$-system and the $GI/G/s$-system are to be compared. It is plausible to apply a *comparison method* as in Example (6.4.c). If we use the *same* seeds $x_0^{(1)}, \ldots, x_0^{(s+1)}$, ($x_0^{(s+1)}$ is not necessary for the $GI/G/(s-1)$-system) for the simulation of both systems, then we obtain exactly the same time distribution for the arrivals and the same sequence of service times for the first $s - 1$ servers. One thus compares these two systems in great detail under the *same conditions* (the same arrivals, the same service times).

The durations of the regeneration cycles Z_i, Y_i in (15) will not be the same in these two simulations. Either one links the regeneration cycles directly to each other in both simulations, or one starts in both simulations again with new cycles which have the same random number seeds (synchronized cycles). This, too, can be described as a question of planning of experiments. $\qquad \square$

Finally let us derive a *generalization of the strong law of large numbers* for a regenerative system. We consider the Markov chain with the pay-offs $r(i)$. Suppose $N(n)$ is the counting process for the renewal process of the Y_i (see (5)). Then we have

$$\frac{1}{n} \sum_{k=0}^{n} r(X_k) = \frac{\sum_{h=1}^{N(n)} Z_h}{\sum_{h=1}^{N(n)} Y_h + (n - \sum_{h=1}^{N(n)} Y_h)} + \frac{\sum_{k=Y_1+\ldots+Y_{N(n)}}^{n} r(X_k)}{n}. \qquad (16)$$

Since there exist infinitely many renewals (see Section 2.2) for a recurrent renewal process with probability 1, $N(n) \to \infty$ with probability 1. Hence it follows from (10) that with probability 1

$$\lim_{n \to \infty} \frac{\sum_{h=1}^{N(n)} Z_h}{\sum_{h=1}^{N(n)} Y_h} = \lim_{n \to \infty} \frac{\sum_{h=1}^{N(n)} Z_h}{\sum_{h=1}^{N(n)+1} Y_h} = g. \tag{17}$$

The second limit relation follows from the fact that (9) has still to hold with probability 1 if the summation extends to $m + 1$ instead of just to m. Since the first term on the right hand side of (16) is, for all n, bracketed by the two expressions on the left hand side of (17), the first expression on the right hand side of (16) also has to tend to g with probability 1 as $n \to \infty$.

Secondly we have

$$\left| \sum_{k=Y_1+\ldots+Y_{N(n)}}^{n} r(X_k) \right| \leqslant \sum_{k=Y_1+\ldots+Y_{N(n)}}^{Y_1+\ldots+Y_{N(n)+1}-1} |r(X_k)|. \tag{18}$$

Let the sum on the right hand side for $h = N(n) + 1$ be denoted by Z_h', in analogy to (4). The Z_h' are finite for all h with probability 1, therefore we have $Z_h'/n \to 0$ as $n \to \infty$ for all h and it follows from this that $Z_{N(n)+1}'/n \to 0$ as $n \to \infty$ with probability 1. From this it follows furthermore that the second expression on the right hand side of (16) vanishes as $n \to \infty$. We have thus with probability 1

$$\lim_{n \to \infty} \frac{1}{n} \sum_{k=0}^{n} r(X_k) = g. \tag{19}$$

(19) has the form of a strong law of large numbers (see (1.4.20)), however, the summands $r(X_k)$ in the arithmetic mean are *not mutually independent*. According to (19) we can therefore also estimate the profit rate by simulating a sufficiently large number n of transitions of the Markov chain and by taking the arithmetic mean (19) as an estimate for g. Recall that (19) holds for chains with an aperiodic recurrent class. Similarly we can conclude that in the $GI/G/s$-system we have

$$\lim_{t \to \infty} \frac{1}{t} \int_0^t X_t \, dt = L, \qquad \lim_{n \to \infty} \frac{1}{n} \sum_{i=1}^{n} W_i = W \tag{20}$$

with probability 1 and that analogous results can be obtained for arbitrary regenerative systems.

Notes to Chapter 6

Random numbers and simulation were used very early in mathematical statistics. But only with the advent of computers did the method become widespread. Generation of canonical random numbers is a central problem of the Monte-Carlo method. [7], [9], [10], [11] discuss this problem extensively and also contain both empirical and analytical tests to assess the statistical quality of random numbers.

The generation of random numbers according to a given distribution by transformation of canonical random numbers is systematically treated in [4] where also the most important special distributions are discussed. [5] shows nicely how to use random numbers.

The problem of simulation of stochastic processes leads finally to questions of model building and programming of a suitable computer program. There are special programming languages for simulations; see [1], [4].

The mathematical foundations of the Monte-Carlo method, especially the variance reduction techniques, are usually developed in the context of computation of integrals, [3], [6]. Statistical techniques, useful for the Monte-Carlo analysis of more complex problems, are collected in [8]. For the regenerative method we refer to [2].

The predominant field of application of stochastic simulation is queueing systems. Also important are risk analysis of investment decisions and network-planning problems. This is by no means an exhaustive list of the applications of simulation. Virtually every stochastic problem which defies analysis by analytical means can be studied by the Monte-Carlo method and simulation.

7

Semi-Markov models

7.1. Markov renewal and semi-Markov processes

In this chapter a useful generalization of both renewal and Markov processes will be considered. For this purpose a new look at continuous parameter Markov chains will be taken. Let $T_n, n = 0, 1, 2, \ldots$, be the times of the nth jumps or transitions of the chain ($T_0 = 0$ the start of the process). Let X_n be the new state of the chain just after the nth jumps. As pointed out at the end of Section 3.3, X_n is a discrete parameter Markov chain with transition probabilities $p_{ij} = q_{ij}/q_i$ for $i \neq j$ and $p_{ii} = 0$. The sojourn times $T_{n+1} - T_n$ in state $X_n = i$ are exponentially distributed with parameter q_i and independent from each other. Furthermore, the following relation holds:

$$P[X_n = j, T_n - T_{n-1} \leqslant t \mid X_0 = i_0, \ldots, X_{n-1} = i_{n-1}, X_n = i;$$

$$T_0 = 0, T_1 \leqslant t_1, \ldots, T_{n-1} \leqslant t_{n-1}]$$

$$= P[X_n = j, T_n - T_{n-1} \leqslant t \mid X_{n-1} = i]$$

$$= p_{ij}(1 - e^{-q_i t}). \tag{1}$$

The first part of this relation will now be taken as a specifying relation of a much broader class of processes, where the sojourn times $T_n - T_{n-1}$ will no longer be restricted to exponential distributions, but may be arbitrarily distributed.

Let I be a denumerable set, X_n, for $n = 0, 1, 2, \ldots$, a family of random variables taking values in I, and $T_n, n = 0, 1, 2, \ldots$, a family of real valued random variables such that $0 = T_0 \leqslant T_1 \leqslant T_2 \leqslant \ldots$. Let $i_0, i_1, \ldots, i_n$ be arbitrary elements of $I, 0 \leqslant t_1 \leqslant t_2 \leqslant \ldots \leqslant t_{n-1}$ and $t \geqslant 0$. If always

$$P[X_n = i_n, T_n - T_{n-1} \leqslant t \mid X_0 = i_0, \ldots, X_{n-1} = i_{n-1};$$

$$T_0 = 0, T_1 \leqslant t_1, \ldots, T_{n-1} \leqslant t_{n-1}]$$

$$= P[X_n = i_n, T_n - T_{n-1} \leqslant t \mid X_{n-1} = i_{n-1}] \tag{2}$$

holds, then the *stochastic process* $\{X_n, T_n, n = 0, 1, 2, \ldots\}$ will be called a *Markov renewal process*. I is called the *state space* of the process and its elements are called *states*. If at a time t, $T_n \leqslant t < T_{n+1}$ we have $X_n = i$, we say the process is in state i at time t. More precisely, a second process $\{Y_t, 0 \leqslant t < L\}$, associated with a Markov renewal process $\{X_n, T_n, n = 0, 1, 2, \ldots\}$, may be defined by

$$Y_t = X_n, \quad \text{if } T_n \leqslant t < T_{n+1}. \tag{3}$$

This process is defined for

$$t < \sup_n T_n = L. \tag{4}$$

L may be finite or infinite (see Example $(6.3.e)$ for a finite L in the case of a continuous Markov chain). $\{Y_t, 0 \leqslant t < L\}$ is called a *semi-Markov process*.

It will be assumed that the right hand side of (2) does not depend on n,

$$P[X_n = j, T_n - T_{n-1} \leqslant t \mid X_{n-1} = i] = Q_{ij}(t). \tag{5}$$

This is the case of a *homogeneous* Markov renewal process. The matrix $Q(t) = (Q_{ij}(t)); i, j \in I$, is called the *semi-Markov kernel* of the Markov renewal process $\{X_n, T_n, n = 0, 1, 2, \ldots\}$. Of course $Q_{ij}(t) = 0$ for $t < 0$. In order to avoid annoying but uninteresting phenomena, it is assumed that $Q_{ij}(0) = 0$. It follows from (5)

$$Q_{ij}(t_1) \leqslant Q_{ij}(t_2) \quad \text{for } t_1 < t_2,$$

$$\lim_{t' \downarrow t} Q_{ij}(t') = Q_{ij}(t),$$

the limit

$$\lim_{t \to \infty} Q_{ij}(t) = p_{ij} \geqslant 0 \tag{6}$$

exists and

$$\sum_{j \in I} p_{ij} = \sum_{j \in I} P[X_n = j \mid X_{n-1} = i] = 1. \tag{7}$$

Hence in general $Q_{ij}(\infty) < 1$ and $Q_{ij}(t)$ are *incomplete* distribution functions.

Theorem 1
$\{X_n, n = 0, 1, 2, \ldots\}$ is a *discrete parameter Markov chain* with transition matrix $P = (p_{ij}); i, j \in I$.

Proof. Using (2) it follows that

$$P[X_n = j \mid X_0 = i_0, \ldots, X_{n-2} = i_{n-2}, X_{n-1} = i]$$

$$= \lim_{t \to \infty} P[X_n = j, T_n - T_{n-1} \leqslant t \mid X_0 = i_0, \ldots, X_{n-1} = i;$$

$$T_0 = 0, T_1 < \infty, \ldots, T_{n-1} \leqslant \infty]$$

$$= \lim_{t \to \infty} P[X_n = j, T_n - T_{n-1} \leqslant t \mid X_{n-1} = i]$$

$$= P[X_n = j | X_{n-1} = i] = p_{ij}.$$

This is equivalent to the *Markov property* (3.1.1). ∎

Next consider the sojourn times in a state i,

$$P[T_n - T_{n-1} \leqslant t | X_n = j, X_{n-1} = i]$$

$$= \frac{P[X_n = j, T_n - T_{n-1} \leqslant t | X_{n-1} = i]}{P[X_n = j | X_{n-1} = i]}$$

$$= Q_{ij}(t)/p_{ij} = F_{ij}(t), \quad t \geqslant 0, \qquad \qquad (8)$$

if $p_{ij} \neq 0$. If $p_{ij} = 0$ then $Q_{ij}(t) = 0$ for all t. In this case define $F_{ij}(t) = 1$ for all $t \geqslant 0$. $F_{ij}(t)$ is the distribution function of the *conditional sojourn time* in state i given that afterwards a transition to state j occurs.

For the unconditional distribution of sojourn times in a state i it follows that

$$P[T_n - T_{n-1} \leqslant t | X_{n-1} = i]$$

$$= \sum_{j \in I} P[X_n = j | X_{n-1} = i] P[T_n - T_{n-1} \leqslant t | X_{n-1} = i, X_n = j]$$

$$= \sum_{j \in I} p_{ij} F_{ij}(t) = \sum_{j \in I} Q_{ij}(t) = F_i(t). \qquad \qquad (9)$$

$F_i(t)$ is the distribution of sojourn time in a state i.

Theorem 2

Let n be an integer $\geqslant 1$; $t_1, \ldots, t_n \in [0, \infty]$ and $i_0, i_1, \ldots, i_n \in I$.

Then

$$P[T_1 - T_0 \leqslant t_1, \ldots, T_n - T_{n-1} \leqslant t_n | X_0 = i_0, \ldots, X_n = i_n]$$

$$= \prod_{\nu=1}^{n} F_{i_{\nu-1} i_\nu}(t_\nu). \qquad \qquad (10)$$

Proof. (10) is valid for $n = 1$. For $n > 1$ (10) is proved by induction. Suppose (10) is valid for n. Then (using Theorem 1)

$$P[T_1 - T_0 \leqslant t_1, \ldots, T_{n+1} - T_n \leqslant t_{n+1} | X_0 = i_0, \ldots, X_{n+1} = i_{n+1}]$$

$$= P[T_{n+1} - T_n \leqslant t_{n+1} | X_0 = i_0, \ldots, X_{n+1} = i_{n+1};$$

$$T_1 - T_0 \leqslant t_1, \ldots, T_n - T_{n-1} \leqslant t_n]$$

$$P[T_1 - T_0 \leqslant t_1, \ldots, T_n - T_{n-1} \leqslant t_n | X_0 = i_0, \ldots, X_{n+1} = i_{n+1}]$$

$$= F_{i_n i_{n+1}}(t_{n+1})$$

$$\times \frac{P[X_{n+1} = i_{n+1}, T_1 - T_0 \leqslant t_1, \ldots, T_n - T_{n-1} \leqslant t_n | X_0 = i_0, \ldots, X_n = i_n]}{P[X_{n+1} = i_{n+1} | X_0 = i_0, \ldots, X_n = i_n]}$$

$$= F_{i_n i_{n+1}}(t_{n+1})$$

$$\times \frac{P[X_{n+1} = i_{n+1} | X_n = i_n] P[T_1 - T_0 \leqslant t_1, \ldots, T_n - T_{n-1} \leqslant t_n | X_0 = i_0, \ldots, X_n = i_n]}{P[X_{n+1} = i_{n+1} | X_n = i_n]}$$

$$= F_{i_n i_{n+1}}(t_{n+1}) \prod_{\nu=1}^{n} F_{i_{\nu-1} i_\nu}(t_\nu).$$

Thus (10) is also valid for $n + 1$.   ∎

Theorem 2 states that, given a sample function of the Markov chain $\{X_n, n = 0, 1, 2, \ldots\}$ the sojourn times in states are independent. Theorems 1 and 2 together explain the name Markov renewal process.

Examples

(a) *Simulation of Markov renewal processes.* Theorems 1 and 2 show how to simulate general Markov renewal processes. Suppose that at some time t the process has jumped to a state i. Then by Theorem 1 the *next* state is simulated by sampling from $p_{ij}, j \in I$. The sojourn time τ is by Theorem 2 and (8) obtained by sampling from the distribution $F_{ij}(t)$. The next jump will occur then at time $t + \tau$ and the whole procedure may be repeated, etc..   □

(b) *Renewal processes.* Let $\{X_n, T_n, n = 0, 1, 2, \ldots\}$ be a Markov renewal process where the state space I contains a single element. Then by Theorem 2 $\{T_n, n = 0, 1, 2, \ldots\}$ is a (non-delayed) *renewal process.*

Let on the other hand $\{S_n, n = 0, 1, 2, \ldots\}$ be a renewal process with some *renewal distribution* $F(t)$. The process may be delayed with the initial distribution $G(t)$. Then it is possible to represent the associated *counting process* $N(t)$ as a *semi-Markov process.* In fact define

$$P[X_1 = 1, T_1 - T_0 \leqslant t | X_0 = 0] = G(t),$$

and
$$P[X_n = n, T_n - T_{n-1} \leqslant t | X_{n-1} = n - 1] = F(t)$$

$$P[X_n = j, T_n - T_{n-1} \leqslant t | X_{n-1} = i] = 0$$

in all other cases. Then $N(t)$ is the semi-Markov process associated with this Markov renewal process and $\{T_n, n = 0, 1, 2, \ldots\}$ corresponds to the renewal process $\{S_n, n = 0, 1, 2, \ldots\}$.   □

(c) *Markov chains.* Both discrete and continuous parameter Markov chains are special cases of semi-Markov processes. In the case of a continuous parameter Markov chain the Markov renewal kernel of the corresponding Markov renewal process is by (1) $Q_{ij}(t) = q_{ij}(1 - e^{-q_i t})/q_i$, for $i \neq j$ and $Q_{ii}(t) = 0$.

A discrete parameter Markov chain $\{X_n, n = 0, 1, 2, \ldots\}$ with transition matrix P can be considered as a Markov renewal process $\{X_n, T_n, n = 0, 1, 2, \ldots\}$ with $T_n = n$. Its kernel is $Q(t) = P\delta(t)$, where $\delta(t) = 0$ for $t < 1$ and $\delta(t) = 1$ for $t \geqslant 1$. Of course no advantage can be gained for the analysis of Markov chains by considering them as Markov renewal processes.   □

(*d*) *Alternating renewal process.* Suppose a piece of equipment has a failure-free time with distribution $L(t)$. After a breakdown the repair lasts a time with distribution $R(t)$. After repair the piece is as new and has again a failure-free period with distribution $L(t)$, etc.. So the piece is alternating in down-state (0) and up-state (1). Let T_0, T_1, T_2 be the successive times of state changes and X_n the simple periodic 2-state chain with transition probabilities $p_{01} = p_{10} = 1, p_{00} = p_{11} = 0$. Then $\{X_n, T_n, n = 0, 1, 2, \ldots\}$ is a Markov renewal process with kernel

and
$$Q_{01}(t) = R(t); \quad Q_{10}(t) = L(t); \quad Q_{00}(t) = Q_{11}(t) = 0;$$
$$F_{01}(t) = R(t); \quad F_{10}(t) = L(t). \qquad \square$$

(*e*) *Availability of a series system.* Suppose a system is built up of m components and the system is functioning only if all the m components are. The distribution of the failure-free interval of component i is assumed to be *exponentially* distributed with parameter λ_i. On failure of a component i the operation of the system is interrupted and no other component fails during the repair of the component. The repair time of component i has distribution $R_i(t)$.

Let $T_n, n = 0, 1, 2, \ldots$, be the successive alternating times of failure and end of repair, let X_n be the component that fails if T_n is a failure time and let $X_n = 0$ if T_n marks the end of a repair. Then $\{X_n, T_n, n = 0, 1, 2, \ldots\}$ is a Markov renewal process. It is

$$Q_{i0}(t) = R_i(t), \quad i \neq 0.$$

Suppose that at some time the system starts functioning again after the end of a repair. Denote by τ_i the failure-free time of component i. Then

$$\begin{aligned}
Q_{0i}(t) &= P[X_n = i, T_n - T_{n-1} \leqslant t | X_{n-1} = 0] \\
&= P[\tau_i \leqslant \tau_j \text{ for all } j \neq i, \tau_i \leqslant t] \\
&= \int_0^t \lambda_i e^{-\lambda_i x} \prod_{j \neq i} e^{-\lambda_j x} \, dx \\
&= \frac{\lambda_i}{\lambda} (1 - e^{-\lambda t}), \quad i \neq 0, \text{ with } \lambda = \sum_{i=1}^m \lambda_i.
\end{aligned}$$

All other $Q_{ij}(t) = 0$. It follows that $p_{i0} = Q_{i0}(\infty) = 1$ for $i > 0$ and $p_{0i} = Q_{0i}(\infty) = \lambda_i/\lambda$ for $i \neq 0$. $\qquad \square$

(*f*) *Reliability of k-out-of-n systems with repair.* Consider the system of example (3.5.*a*), but suppose that the repair-time distribution is no longer exponential, but is an arbitrary distribution $R(t)$. Of course now the system can no longer be modelled as a Markov chain. But in the case of a *single* repair channel, it can be treated as a Markov renewal process. Let $T_0, T_1, T_2, \ldots$ be the instants of termination of a repair and X_n the number of components still to be repaired at the termination of the nth repair. Then $\{X_n, T_n, n = 0, 1, 2, \ldots\}$ is a Markov renewal process. In order to determine its kernel we have to compute the distribution of new component failures during a repair. These failures are generated by a *birth process* with $\lambda_i = k\lambda$ (cold reserve) or $\lambda_i = (n - i)\lambda$ (hot reserve) for $i \leqslant n - k$, $\lambda_i = 0$ for $i > n - k$ (this assumes that operation is interrupted after system breakdown, but other assumptions would also be possible). Let $p_{ij}(t)$ be the transition probabilities of this birth process (of course $p_{ij}(t) = 0$ for $i > j$).

Now $Q_{ij}(t) = 0$ for $j < i - 1$ because $j - i + 1$, the number of components that fail during a repair, must be non-negative. For $j \geqslant i - 1$, $Q_{ij}(t)$ is the probability that the repair lasts at most t and that $j - i + 1$ components fail during repair, given that i components have failed at the start. Thus for $i > 0$

$$Q_{ij}(t) = \int_0^t p_{ij+1}(x)\,\mathrm{d}R(x).$$

For $i = 0$ there will be an interval to the first failure and only then will repair begin. Thus in this case

$$Q_{0j}(t) = \int_0^t k\lambda e^{-\lambda k x} Q_{1j}(t - x)\,\mathrm{d}x$$

(in the case of cold reserve, for hot reserve $k\lambda$ has to be replaced by $n\lambda$).   □

(g) *M/G/1-queueing system.* This system has been treated in Section 4.2. by means of an embedded Markov chain X_n. X_n is the number of customers left behind immediately after a departure. Let $T_n, n = 0, 1, 2, \ldots$, be the times of these departures. Then $\{X_n, T_n, n = 0, 1, 2, \ldots\}$ is a Markov renewal process. $Q_{ij}(t)$ is for $i > 0$ the probability that service lasts at most t and that during the service $j - i + 1$ clients arrive. Denote this probability by $q_{j-i+1}(t)$; thus $Q_{ij}(t) = q_{j-i+1}(t), j \geqslant i - 1$. Of course $Q_{ij}(t) = 0$ for $j < i - 1$. As the arrivals form a *Poisson process* it follows that

$$q_n(t) = \int_0^t \frac{(\lambda x)^n}{n!} e^{-\lambda x}\,\mathrm{d}G(x), \quad n = 0, 1, \ldots$$

if λ is the arrival rate and $G(t)$ the service-time distribution. For $i = 0$ service begins only after the next arrival. Therefore

$$Q_{0j}(t) = \int_0^t \lambda e^{-\lambda x} q_j(t - x)\,\mathrm{d}x.$$

Verify that $p_{ij} = Q_{ij}(\infty)$ correspond to the transition probabilities (4.2.4) and (4.2.5) of the embedded Markov chain.   □

(h) *GI/M/1-queueing system.* According to Section 4.1 this is a one-server queueing system where the arrivals form a *renewal process* with a renewal distribution $G(t)$ which is the distribution of the inter-arrival times. The

service time is exponentially distributed with a parameter μ. Let $T_n, n = 0, 1,$
$2, \ldots$, be the times of arrivals and X_n be the number of customers in the
system just before the nth arrival. Then $\{X_n, T_n, n = 0, 1, 2, \ldots\}$ is a Markov
renewal process.

For $j \leqslant i + 1$, $Q_{ij}(t)$ is the probability that the time to the next arrival is at
most t and that during this inter-arrival time $i + 1 - j$ customers are served.
For $j > 0$ the service process is just a Poisson arrival process as in Example
(g). Hence

$$Q_{ij}(t) = q_{i+1-j}(t), \quad i = 0, 1, 2, \ldots; 0 < j \leqslant i + 1,$$

where $q_n(t)$ is defined as in Example (g) with μ in place of λ. For $j = 0$ all
$i + 1$ customers present at the beginning of the interval are served during the
interval, i.e. more than i customers are served. Thus

$$Q_{i0}(t) = \int_0^t \left\{ 1 - \sum_{k=0}^i \frac{(\mu x)^k}{k!} e^{-\mu x} \right\} dG(x)$$

$$= G(t) - \sum_{n=0}^i q_n(t).$$

Of course $Q_{ij}(t) = 0$ for $j > i + 1$. $\qquad\qquad \Box$

7.2. Limiting behaviour of regenerative processes

As with Markov chains, we are interested in the limiting behaviour of
a semi-Markov process for $t \to \infty$, especially in the limits of the state probabilities $P[Y_t = i]$ as $t \to \infty$. These limits may be obtained using *renewal
theory*. It turns out that such limits may be obtained for even more general
processes than semi-Markov processes.

Let $\{X_n, T_n, n = 0, 1, 2, \ldots\}$ be a Markov renewal process. If $X_0 = i \in I$ let
$S_m^{ij}, m = 1, 2, \ldots$, be the successive T_n for which $X_n = j$ and $S_0^{ij} = 0$. Then
$\{S_m^{ij}, m = 0, 1, 2, \ldots\}$ form *renewal processes* which are delayed if $i \neq j$. Let

$$G_{ij}(t) = P[S_1^{ij} \leqslant t] \qquad\qquad (1)$$

be the distribution of the *first passage time to state j starting from state i* of
the Markov renewal process. Then $G_{ij}(t)$ is the *initial distribution* of the
renewal process $\{S_m^{ij}, m = 0, 1, 2, \ldots\}$ and $G_{jj}(t)$ is its *renewal distribution*.

State $j \in I$ is called *recurrent (transient)* if its associated renewal processes
$\{S_m^{ij}, m = 0, 1, 2, \ldots\}$ are recurrent (transient), i.e. if $G_{jj}(\infty) = 1$ $(G_{jj}(\infty) < 1)$.
From this it follows immediately that $j \in I$ is recurrent (transient) if and
only if it is recurrent (transient) in the *embedded Markov chain*
$\{X_n, n = 0, 1, 2, \ldots\}$.

Now define the transition probability functions of the semi-Markov process Y_t (7.1.3)

$$P[Y_t = j | X_0 = i] = p_{ij}(t). \qquad\qquad (2)$$

State i *leads* to state j, $i \Rightarrow j$, if there exists a $t > 0$ such that $p_{ij}(t) > 0$. i and j *communicate*, $i \Leftrightarrow j$, if $i \Rightarrow j$ and $j \Rightarrow i$. It is evident that i leads to or communicates with j if and only if it does so in the embedded Markov chain $\{X_n, n = 0, 1, 2, \ldots\}$. Hence the whole classification of states into disjoint classes can be carried over from X_n to the semi-Markov process Y_t (see Section 3.2). In particular the notion of *irreducibility* carries over and also Theorem 3.2.5 which states that the states of a class are either all recurrent or all transient. A warning should be given however that the notion of positive and null recurrence do not necessarily carry over from X_n to Y_t.

A state j is called *periodic* with *period* λ if $G_{jj}(t)$ is *arithmetic* with jumps only at $t = h\lambda$, $h = 1, 2, \ldots$. If $G_{jj}(t)$ is non-arithmetic, j is called *aperiodic*. Again the periodicity of a state j in the semi-Markov process has nothing to do with its possible periodicity in the embedded Markov chain. The latter may well be periodic and Y_t is still aperiodic (see Example $(7.1.d)$). But there is an analogue to Theorem 3.2.3.

Theorem 1
Either all states of a class have the same period or all are aperiodic.

Proof. Let $i \Leftrightarrow j$. Considering in the original Markov renewal process only the times T_n for which $X_n = i$ or $X_n = j$, one obtains a new Markov renewal process with a kernel

$$\hat{Q} = \begin{bmatrix} \hat{Q}_{ii} & \hat{Q}_{ij} \\ \hat{Q}_{ji} & \hat{Q}_{jj} \end{bmatrix}.$$

It is evident that in this new process one has the same renewal processes S_m^{ii}, S_m^{ij}, S_m^{ji}, S_m^{jj} as in the original process. For the first passage times it follows that

$$\begin{aligned} G_{ii} &= \hat{Q}_{ii} + \hat{Q}_{ij} * \hat{Q}_{ji} + \hat{Q}_{ij} * \hat{Q}_{jj} * \hat{Q}_{ji} + \ldots \\ &\quad + \hat{Q}_{ij} * \hat{Q}_{jj}^{*n} * \hat{Q}_{ji} + \ldots, \end{aligned} \tag{3}$$

$$\begin{aligned} G_{jj} &= \hat{Q}_{jj} + \hat{Q}_{ji} * \hat{Q}_{ij} + \hat{Q}_{ji} * \hat{Q}_{ii} * \hat{Q}_{ji} + \ldots \\ &\quad + \hat{Q}_{ji} * \hat{Q}_{ii}^{*n} * \hat{Q}_{ij} + \ldots . \end{aligned} \tag{4}$$

If i is periodic with period λ then all points of increase of G_{ii} are contained in $J = \{\lambda, 2\lambda, 3\lambda, \ldots\}$ and this must also be true for the terms of (3). A point of increase of a term $\hat{Q}_{ij} * \hat{Q}_{jj}^{*n} * \hat{Q}_{ji}$ is of the form $c = a + b$, where a is a point of increase of $\hat{Q}_{ij} * \hat{Q}_{ji}$ and b of $\hat{Q}_{jj}^{*n}$. Since $c, a \in J$ it follows that $b \in J$ and hence the points of increase of (4) are also in J. This implies that the period of j is an integer multiple of λ. Repeating the argument with reversed roles for i and j proves the first part of the theorem. But if i is aperiodic then j cannot be periodic, because otherwise i would have the same period as j. ∎

In the sequel only *aperiodic* classes of states will be considered in order to limit the discussion. It is left to the reader to obtain the corresponding results for the periodic cases. The following theorem establishes the existence of the limits of $p_{ij}(t)$ as $t \to \infty$. Note that this theorem is an extension of Theorem 3.4.1 and that the subsequent proof would also have been an alternative proof of Theorem 3.4.1.

Theorem 2
(a) If $j \in I$ is *transient*, then for all $i \in I$

$$\lim_{t \to \infty} p_{ij}(t) = 0. \tag{5}$$

(b) If $j \in I$ is *recurrent* and the *mean passage time*

$$\mu_{jj} = \int_0^\infty (1 - G_{jj}(t)) dt < \infty,$$

then

$$\lim_{t \to \infty} p_{ij}(t) = G_{ij}(\infty) \frac{\mu_j}{\mu_{jj}}, \tag{6}$$

where

$$\mu_j = \int_0^\infty (1 - F_j(t)) dt$$

is the *mean sojourn time* in state j.

Proof. If $X_0 = j$, then $Y_t = j$ if either no transition has occurred up to t ($T_1 - T_0 > t$) or Y_t has returned for the first time to j at some time $x < t$ and the process starting afresh at time x is in state j at time $t - x$. Hence, by this renewal argument,

$$p_{jj}(t) = (1 - F_j(t)) + \int_0^t p_{jj}(t - x) dG_{jj}(x). \tag{7}$$

As $1 - F_j(t) \leqslant 1 - G_{jj}(t)$ and $\mu_{jj} < \infty$ the function $1 - F_j(t)$ satisfies the assumptions of Theorem 2.3.4 if j is recurrent and it follows that $\mu_j \leqslant \mu_{jj} < \infty$ and

$$\lim_{t \to \infty} p_{jj}(t) = \mu_j / \mu_{jj}. \tag{8}$$

Because in this case $G_{jj}(\infty) = 1$ this proves (6) for $i = j$. If j is transient, then (5) follows for $i = j$ from (2.3.1).

Again by a renewal argument for $i \neq j$

$$p_{ij}(t) = \int_0^t p_{jj}(t - x) dG_{ij}(x).$$

From this and (8), (5) follows and also (6) for $i \neq j$.   ∎

If $i \leftrightarrow j$ and i and j are recurrent, then by Theorem 3.2.8 j is reached from i with probability 1. This implies $G_{ij}(\infty) = 1$. Hence the following corollary holds.

Corollary
For all states i, j of a *recurrent class*, if $\mu_{jj} < \infty$

$$\lim_{t \to \infty} p_{ij}(t) = \mu_j / \mu_{jj}. \tag{9}$$

This result has an obvious intuitive interpretation!

In order to compute μ_{jj}, it is sufficient to consider irreducible classes and each such class can be considered independently from all the others. Thus the following theorem can be applied to each irreducible class of a semi-Markov process.

Theorem 3
Suppose the Markov renewal process $\{X_n, T_n, n = 0, 1, 2, \ldots\}$ is *irreducible* and *recurrent*, such that the embedded Markov chain $\{X_n, n = 0, 1, 2, \ldots\}$ is *positive recurrent*. Let $u_i > 0$, $i \in I$ be a solution of

$$u_j = \sum_{i \in I} u_i p_{ij}, \quad j \in I. \tag{10}$$

Then

$$\mu_{jj} = \frac{1}{u_j} \sum_{i \in I} u_i \mu_i. \tag{11}$$

Proof. Define

$$\mu_{ij} = \int_0^\infty (1 - G_{ij}(t)) \, dt \tag{12}$$

as the mean first passage time from i to j and

$$m_{ij} = \int_0^\infty (1 - F_{ij}(t)) \, dt \tag{13}$$

as the mean conditional sojourn time in i given that the next transition is to j. Then

$$\begin{aligned}
\mu_{ij} &= p_{ij} m_{ij} + \sum_{k \neq j} p_{ik}(m_{ik} + \mu_{kj}) \\
&= \sum_k p_{ik} m_{ik} + \sum_{k \neq j} p_{ik} \mu_{kj} \\
&= \mu_i + \sum_{k \neq j} p_{ik} \mu_{kj}.
\end{aligned} \tag{14}$$

According to Theorem 3.2.9 a bounded solution $u_i > 0$ to (10) exists (e.g. the stationary distribution of X_n) and is unique up to a multiplicative constant. Multiplying both sides of (14) by u_i and summing over i gives

$$\sum_i u_i \mu_{ij} = \sum_i u_i \mu_i + \sum_{k \neq j} \mu_{kj} \sum_i u_i p_{ik}$$

$$= \sum_i u_i \mu_i + \sum_{k \neq j} u_k \mu_{kj},$$

from which

$$u_j \mu_{jj} = \sum_i u_i \mu_i.$$

This corresponds to (11). ∎

Note that the mean passage times μ_{jj} of the states can be computed using only transition probabilities of the embedded Markov chain and mean sojourn times μ_i in the states i.

By the corollary to Theorem 2 this holds also for the limiting probabilities $p_{ij}(\infty)$.

Theorem 4
Under the assumptions of Theorem 3

$$\lim_{t \to \infty} p_{ij}(t) = u_j \mu_j \bigg/ \sum_{i \in I} u_i \mu_i. \tag{15}$$

Proof. (15) follows from (9) and (11). ∎

The following examples give some applications of these results.

Examples
(a) *Alternating renewal process.* The process of Example (7.1.d) satisfies the assumptions of Theorems 3 and 4 with $u_0 = u_1 = 1$ (for example). μ_0 and μ_1 are the mean repair time R and failure-free time L respectively. Thus by (15), we have

$$\lim_{t \to \infty} p_{i0} = p_0 = \frac{R}{R + L}, \quad i = 0, 1,$$

$$\lim_{t \to \infty} p_{i1} = p_1 = \frac{L}{R + L}, \quad i = 0, 1,$$

which correspond to intuition. (11) gives $\mu_{00} = \mu_{11} = R + L$ (which is obvious on first grounds). □

(b) *Availability of a series system.* Theorems 3 and 4 may also be applied to Example (7.1.e). One may choose $u_i = \lambda_i/\lambda$ for $i = 1, \ldots, m$; $u_0 = 1$. For

$i \neq 0$, μ_i is the mean repair time R_i of component i. Because $F_0(t) = Q_{01}(t) + \ldots + Q_{0m}(t) = 1 - e^{-\lambda t}$ (see Example $(7.1.e)$) it follows that $\mu_0 = 1/\lambda$. Hence by (11) the mean time between two restarts of operation is

$$\mu_{00} = \frac{\lambda_1 R_1 + \lambda_2 R_2 + \ldots + \lambda_m R_m + 1}{\lambda},$$

and the mean times between two consecutive failures of component j are

$$\mu_{jj} = \frac{1}{\lambda_j} + \sum_{i=1}^{m} \frac{\lambda_i}{\lambda_j} R_i.$$

For the limiting probabilities $p_j = \lim_{t \to \infty} p_{ij}(t)$ we obtain from (15)

$$p_0 = (1 + \lambda_1 R_1 + \ldots + \lambda_m R_m)^{-1},$$

$$p_j = \lambda_j R_j p_0, \quad j = 1, 2, \ldots, m.$$

Note that for $m = 1$ the results of Example (a) are obtained (with $L = 1/\lambda$). □

(c) *M/G/1-queueing system.* In Example $(7.1.g)$ the event $[Y_t = j]$ signifies that after the last departure of a customer before time t there were j customers left behind to serve. So

$$p_j = \lim_{t \to \infty} p_{ij}(t)$$

is the probability of this event for large t. This should not be confused with the probability that at some time t there are j customers in the system.

According to Section 4.2 the embedded Markov chain is *transient* if the expected number of arrivals r during a service time is greater than 1. In this case $p_j = 0$ for all $j = 0, 1, 2, \ldots$. Otherwise the process is *recurrent* and if $r < 1$ there exists a stationary distribution u_j of the embedded Markov chain (the chain is positive recurrent). The mean sojourn times in states $j \neq 0$ are $\mu_j = S$, where S is the mean service time. For the mean sojourn time in state 0 we have to add the mean time $1/\lambda$ till the next arrival and so $\mu_0 = 1/\lambda + S$. Hence for $r < 1$ Theorem 3 applies and it follows from (11), for instance, that

$$\mu_{00} = 1/\lambda + S/u_0 = 1/\lambda + S/(1 - r),$$

because $u_0 = 1 - r$ (see $(4.2.12)$). A *busy period* for the server is the time span from the arrival of a customer finding the server idle till the next time the server becomes idle. μ_{00} evidently is the sum of the mean idle time $1/\lambda$ plus the mean busy period. Hence the latter equals $S/(1 - r)$. □

(d) *Mean first passage times in Markov chains.* (14) is a system of linear equations which determines the mean first passage times μ_{ij}. In the case of a discrete parameter Markov chain $\mu_i = 1$ and (14) becomes

$$\mu_{ij} = 1 + \sum_{k \neq j} p_{ik}\mu_{kj}.$$

For a continuous parameter Markov chain $\mu_i = 1/q_i$ and $p_{ik} = q_{ik}/q_i$ for $k \neq i, p_{ii} = 0$, and therefore

$$q_i\mu_{ij} = 1 + \sum_{\substack{k \neq j \\ k \neq i}} q_{ik}\mu_{kj}.$$

These are useful equations for the analysis of Markov chains. □

For the $M/G/1$-queueing system of Example (7.1.g) we have considered the number of customers in the system only at the instances T_n immediately after a departure. But we may be interested in the number Z_t of customers in the system at an *arbitrary* time t. For the stochastic process $\{Z_t, t \geq 0\}$ the transition probabilities $q_{ij}(t) = P[Z_t = j | X_0 = i]$ may be defined, X_0 being the initial number of customers. Is it possible to establish the existence of the limiting probabilities $\lim_{t \to \infty} q_{ij}(t)$ and to compute them?

In fact Theorem 2 can be extended to cover the case of the process $\{Z_t, t \geq 0\}$ above. This is a striking proof of the power of renewal theory. Let in general $\{Z_t, t \geq 0\}$ be a stochastic process with Z_t being *discrete* random variables taking values in some denumerable *state space I*. Let $p_j(t) = P[Z_t = j]$ for all $j \in I$. Suppose there is a *non-delayed* renewal process $\{S_n, n = 0, 1, 2, \ldots\}$ with renewal distribution $F(t)$. If for $t > s$

$$P[Z_t = j | S_1 = s] = p_j(t - s) \tag{16}$$

then $\{Z_t, t \geq 0\}$ is called a *regenerative process*. (16) means intuitively that at each renewal time $S_0, S_1, S_2, \ldots$ the process starts from scratch; it regenerates itself. Of course semi-Markov processes are regenerative, each one of the embedded renewal processes $\{S_n^{jj}, n = 0, 1, 2, \ldots\}$ defines regeneration points. In order to limit the discussion it will again be supposed that $\{S_n, n = 0, 1, 2, \ldots\}$ is *aperiodic*.

The following theorem extends Theorem 2.

Theorem 5

Let $\{Z_t, t \geq 0\}$ be a regenerative process with a *recurrent* renewal process $\{S_n, n = 0, 1, 2, \ldots\}$.

Define

$$q_j(t) = P[Z_t = j, S_1 > t]. \tag{17}$$

Then for all $j \in I$

$$\lim_{t \to \infty} p_j(t) = \lim_{t \to \infty} P[Z_t = j] = \frac{1}{\mu} \int_0^\infty q_j(t)\,dt, \tag{18}$$

where μ is the mean renewal time

$$\mu = \int_0^\infty (1 - F(t)) \mathrm{d}t < \infty. \tag{19}$$

Proof. The proof parallels the proof of Theorem 2. By a renewal argument using (16) it follows that

$$p_j(t) = q_j(t) + \int_0^t p_j(t - x) \mathrm{d}F(x). \tag{20}$$

Then Theorem 2.3.4 is applicable and implies (18).   ∎

This theorem establishes the existence of the limiting probabilities of regenerative processes. Using further structural properties of the processes under study, one may even be able to compute these limits as the following section shows.

Example
(e) *M/G/1-queueing system.* Let Z_t be as above the number of customers in the system at time t. Let $q_{jj}(t) = P[Z_t = j | X_0 = j]$. Then $\{Z_t, t \geqslant 0\}$ can be considered as a regenerative process with renewal process $\{S_n^{jj}, n = 0, 1, 2, \ldots\}$ (the corresponding embedded renewal process of the Markov renewal process introduced in Example (7.1.g)). If $r < 1$ Theorem 5 applies because $\{S_n^{jj}, n = 0, 1, 2, \ldots\}$ is recurrent (see Example (c) and Section 4.2).   □

7.3. Markov renewal equations

In this section the theory of *renewal equations* (see Section 2.1) will be generalized. This allows the study of Markov renewal processes or semi-Markov processes for finite times t, just as Kolmogorov's equations determine continuous parameter Markov chains. Furthermore this opens a new approach for the determination of the asymptotic behaviour of semi-Markov processes and also of some special regenerative processes.

If $\{X_n, T_n, n = 0, 1, 2, \ldots\}$ is a Markov renewal process, let $N_{ij}(t)$ be the *counting process* associated with the renewal process $\{S_m^{ij}, m = 0, 1, 2, \ldots\}$ (see (2.1.3)). As in Section 2.1 we stress that the point $t = 0$ will not be counted as a renewal in $N_{ij}(t)$. Furthermore let $M_{ij}(t) = E[N_{ij}(t)]$ be the corresponding *renewal function*. The matrix $M(t) = (M_{ij}(t))$, $i, j \in I$, is called a *Markov renewal kernel.*

According to (2.1.6) for $t \geqslant 0$

$$M_{ij}(t) = \sum_{k=1}^\infty G_{ij} * G_{jj}^{*(k-1)}(t), \tag{1}$$

where $G_{jj}^{*0}(t) = 1$ for $t \geqslant 0$ and according to (2.1.7)

$$M_{ij}(t) = G_{ij}(t) + \int_0^t M_{ij}(t-x)\mathrm{d}G_{jj}(x). \tag{2}$$

Now suppose the Markov renewal process starts in state i. What is the probability that there was a first transition to state j before t? Either (i) the first transition takes place before t and goes to j; the probability of this event is $Q_{ij}(t)$; or (ii) a first transition to $k \neq j$ takes place at time $s < t$ and there is a first passage from k to j in a time of at most $t - s$. Thus by this renewal argument it follows that

$$G_{ij}(t) = Q_{ij}(t) + \int_0^t \sum_{k \neq j} G_{kj}(t-s)\mathrm{d}Q_{ik}(s). \tag{3}$$

This implies for $h = 1, 2, \ldots$

$$G_{ij} * G_{jj}^{*(h-1)} = \left\{ Q_{ij} + \sum_{k \neq j} Q_{ik} * G_{kj} \right\} * G_{jj}^{*(h-1)}$$

$$= Q_{ij} * G_{jj}^{*(h-1)} + \sum_{k \neq j} Q_{ik} * (G_{kj} * G_{jj}^{*(h-1)}). \tag{4}$$

Summing this for $h = 1, 2, \ldots$ yields, using (1),

$$M_{ij} = Q_{ij} * (M_{jj} + 1) + \sum_{k \neq j} Q_{ik} * M_{kj}$$

$$= Q_{ij} + \sum_{k \in I} Q_{ik} * M_{kj} = Q_{ij} + \sum_{k \in I} M_{kj} * Q_{ik} \tag{5}$$

because the convolution $Q_{ik} * M_{kj}$ is commutative as is seen from (4). The last equation in (5) has an evident probabilistic interpretation: perhaps a first renewal in state j is due to a first transition to j. This event has probability Q_{ij}. In any case after a first transition to a state $k \in I$ at some time $s < t$, there remains the expected number of $M_{kj}(t - s)$ renewals in state j in the remaining period $t - s$. In matrix notation (5) becomes

$$M = Q + Q * M, \tag{6}$$

where the matrix convolution $Q * M$ is defined in an analogous way to normal matrix multiplication. Suppose $Z(t)$ and $z(t)$ are vectors of functions $(Z_i(t))$ and $(z_i(t))$ for $i \in I$ and defined for $t \geq 0$ respectively, $z(t)$ is given and $Q(t)$ is some *semi-Markov renewal* kernel. Then equations (6) are special cases of the more general equations

$$Z = z + Q * Z, \quad t \geq 0. \tag{7}$$

Systems of equations of the form (7) are called *Markov renewal equations*. Equations (2.1.8) and (2.1.9) can be considered as special cases of (7) and (6) respectively, just as ordinary renewal processes are special Markov renewal processes.

Examples

(a) *Transition probabilities of semi-Markov processes.* Let $\{Y_t, t \geqslant 0\}$ be the semi-Markov process associated with a Markov renewal process $\{X_n, T_n, n = 0, 1, 2, \ldots\}$ with kernel Q. Its transition probabilities $p_{ij}(t)$ are defined by (7.2.2). Now

$$p_{ij}(t) = P[Y_t = j; T_1 > t | X_0 = i] + P[Y_t = j; T_1 \leqslant t | X_0 = i]$$

and $P[Y_t = j; T_1 > t | X_0 = i] = 0$ for $j \neq i$, whereas $P[Y_t = i, T_1 > t | X_0 = i] = 1 - F_i(t)$, where $F_i(t)$ is the distribution of sojourn time in state i; see (7.1.9). Therefore it follows that

$$P[Y_t = j, T_1 > t | X_0 = i] = \delta_{ij}(1 - F_i(t)).$$

Concerning the second term in the above decomposition, a renewal argument referring to the first transition at time $s \leqslant t$ to some state $k \in I$ is used. This leads to

$$p_{ij}(t) = \delta_{ij}(1 - F_i(t)) + \sum_{k \in I} \int_0^t p_{kj}(t - s) \mathrm{d}Q_{ik}(s).$$

For every fixed $j \in I$ this represents a system of Markov renewal equations. □

(b) *Kolmogorov's equations for continuous Markov chains.* In the case of a continuous parameter Markov chain we have $Q_{ij}(t) = q_{ij}(1 - \mathrm{e}^{-q_i t})/q_i$ for $i \neq j$ (see Example (7.1.c)) and $1 - F_i(t) = \mathrm{e}^{-q_i t}$. Hence the Markov renewal equations for the transition probabilities of Example (a) become

$$p_{ij}(t) = \delta_{ij}\mathrm{e}^{-q_i(t)} + \sum_{k \neq i} q_{ik} \int_0^t p_{kj}(t - s)\mathrm{e}^{-q_i s}\mathrm{d}s.$$

These are the equations already found in (3.3.18) and (3.3.19). As noted there, they are equivalent to Kolmogorov's backward equations. The forward equations cannot be derived directly from the Markov renewal equations, but once the solution to these latter equations is found, it can be shown that this solution in fact satisfies the forward equations; see Example (f) below. □

(c) *Functionals of semi-Markov processes.* Let f be a real valued function on the state space I of a Markov renewal process $\{X_n, T_n, n = 0, 1, 2, \ldots\}$. If the corresponding semi-Markov process $\{Y_t, t \geqslant 0\}$ is at some instant t in a state j, then $f(j)$ can be considered as an *earning rate* (or sometimes also a cost rate). $f(Y_s)$ for $0 \leqslant s \leqslant t < L$ (see (7.1.4)) is a stream of earning rates over the period $[0, t]$. At a discount rate $0 \leqslant \rho$ this represents a *present value* (discounted at time 0) of

$$W(t) = \int_0^t \mathrm{e}^{-\rho s}f(Y_s)\mathrm{d}s.$$

$\{W(t), t \geqslant 0\}$ is called a *functional* of the semi-Markov process $\{Y_t, t \geqslant 0\}$.

Let $E_j[W(t)]$ be the *expected value* of $W(t)$ if the semi-Markov process starts in state j. Suppose the first transition takes place at time $T_1 = x$. The present value of the earnings up to x equals (if $\rho > 0$)

$$\int_0^x e^{-\rho s} f(j)\,ds = \frac{f(j)}{\rho}(1 - e^{-\rho x}).$$

With probability $1 - F_j(t)$ there is no transition up to time t. The present value $W(t)$ in this case is given by the above formulae with $x = t$.

Otherwise the first transition takes place at an instant x (with probability $dF_j(x)$) and goes to state k (with probability $dQ_{jk}(x)$). Then the present value till this first transition is given by the above formulae and the expected value of the functional, discounted to 0 in the remaining interval $[x, t]$, equals $e^{-\rho x} E_k[W(t - x)]$. Using this renewal argument one obtains for $\rho > 0$

$$E_j[W(t)] = \frac{f(j)}{\rho}(1 - e^{-\rho t})(1 - F_j(t))$$

$$+ \int_0^t \frac{f(j)}{\rho}(1 - e^{-\rho x})\,dF_j(x)$$

$$+ \sum_{k \in I} \int_0^t E_k[W(t - x)]e^{-\rho x}\,dQ_{jk}(x).$$

These are renewal equations with a renewal kernel $\tilde{Q}$, such that $d\tilde{Q}_{jk}(x) = e^{-\rho x}\,dQ_{jk}(x)$. For $\rho = 0$ the same argument leads to

$$E_j[W(t)] = f(j)t(1 - F_j(t)) + \int_0^t f(j)x\,dF_j(x)$$

$$+ \sum_{k \in I} \int_0^t E_k[W(t - x)]\,dQ_{jk}(x).$$

These are renewal equations with kernel Q. It is worthwhile to specialize these equations for the important case of continuous parameter Markov chains. But this is left to the reader. $\qquad\square$

(*d*) *Embedded Markov renewal processes.* In this example more structured regenerative processes than those at the end of Section 7.2 will be defined. As there, let $\{Z_t, t \geq 0\}$ be a stochastic process with Z_t being discrete random variables taking values in a denumerable state space I. Suppose there exist times $0 = T_0 \leq T_1 \leq T_2 \leq \ldots$ such that, with

$$X_n = Z_{T_n},$$

the process $\{X_n, T_n, n = 0, 1, 2, \ldots\}$ is a Markov renewal process with a kernel Q. Define the transition probabilities $q_{ij}(t) = P[Z_t = j | X_0 = i]$. If furthermore the relation

$$P[Z_t = j | T_1 = s, X_1 = k] = q_{kj}(t - s)$$

holds for all $s < t$ and $k \in I$, then $\{X_n, T_n, n = 0, 1, 2, \ldots\}$ is called an *embedded Markov renewal process* of $\{Z_t, t \geqslant 0\}$. Note that the process Z_t defined in Example (7.2.e) has such an embedded Markov renewal process. Taking as T_n the times of departures $\{X_n, T_n, n = 0, 1, 2, \ldots\}$ is in fact the Markov renewal process introduced in Example (7.1.g).

Let $r_{ij}(t) = P[Z_t = j, T_1 > t | X_0 = i]$. As $[Z_t = j] = [Z_t = j, T_1 > t] \cup \{\cup_{k \in I} [Z_t = j, T_1 \leqslant t, X_1 = k]\}$ it follows, using the above renewal property, that

$$q_{ij}(t) = r_{ij}(t) + \sum_{k \in I} \int_0^t q_{kj}(t - x) \mathrm{d}Q_{ik}(x).$$

These are again Markov renewal equations. □

The next step will be to generalize Theorem 2.1.2 for Markov renewal equations. For this purpose the following representation of the Markov renewal kernel M in terms of the semi-Markov kernel is useful:

Theorem 1
For all $t \geqslant 0$

$$M(t) = \sum_{k=1}^{\infty} Q^{*k}(t). \tag{8}$$

Proof. Fix an initial state $i \in I$ and define the indicator function $I_{j,t}(X_n, T_n) = 1$ if $X_n = j$ and $T_n \leqslant t$, and 0 otherwise. Then

$$N_{ij}(t) = \sum_{n=1}^{\infty} I_{j,t}(X_n, T_n)$$

and

$$M_{ij}(t) = E[N_{ij}(t)] = E\left[\sum_{n=1}^{\infty} I_{j,t}(X_n, T_n)\right]$$

$$= \sum_{n=1}^{\infty} P[X_n = j, T_n \leqslant t | X_0 = i]. \tag{9}$$

Define for an instance $Q_{ij}^{(n)}(t) = P[X_n = j, T_n \leqslant t | X_0 = i]$. Then by an evident renewal argument

$$Q_{ij}^{(n+1)}(t) = \sum_{k \in I} \int_0^t Q_{kj}^{(n)}(t - x) \mathrm{d}Q_{ik}(x)$$

and this shows that $Q^{(n)} = Q^{*n}$. Therefore (8) follows from (9). ■

In order to study the solution of the Markov renewal equations (7), the class of functions $z(t) = (z_i(t))$, $i \in I$ and defined for $t \geqslant 0$, will be limited. It will be assumed, that

(i) $\sup_{i \in I} z_i(t)$ is bounded on every finite interval.
A family of functions $z_i(t), i \in I, t \geqslant 0$, is said to belong to class B if it satisfies (i). If furthermore

(ii) $z_i(t) \geqslant 0$ for $t \geqslant 0$ and all $i \in I$,
the family is said to belong to class B_+. The following theorem is an existence theorem for a solution of the Markov renewal equations (7).

Theorem 2
If $z \in B_+$ and $M * z \in B$, then

$$Z = z + M * z \tag{10}$$

is a solution to the Markov renewal equations (7) and $Z \in B$.

Proof. First we show that $z \in B$ implies $Q * z \in B$. Let

$$\gamma(t) = \sup_{x \leqslant t} \left(\sup_{i \in I} |z_i(x)| \right).$$

Then for all $t \geqslant 0$ and $j \in I$

$$|(Q * z)_j(t)| = \left| \sum_k \int_0^t z_k(t-x) \mathrm{d}Q_{jk}(x) \right|$$

$$\leqslant \sum_k \int_0^t |z_k(t-x)| \mathrm{d}Q_{jk}(x)$$

$$\leqslant \sum_k \int_0^t \gamma(t) \mathrm{d}Q_{jk}(x) = \sum_k Q_{jk}(t)\gamma(t) \leqslant \gamma(t). \tag{11}$$

This shows that $Q * z \in B$.
We have formally, using (6),

$$Q * Z = Q * (z + M * z) = Q * z + Q * (M * z)$$

$$= Q * z + (Q * M) * z = Q * z + (M - Q) * z$$

$$= M * z = Z - z.$$

The assumption that $z, M * z \in B$ and hence $Q * (M * z) \in B$ allows the application of the distributive law for matrix convolution and the further assumption $z \in B_+$ also allows the application of the associative law which holds for non-negative matrices. This shows that (10) is in fact a solution of (7). $Z \in B$ follows from $z, M * z \in B$. ∎

Theorem 2 generalizes the existence part of Theorem 2.1.2. In fact we have $M * z \in B$, if I is a finite state space. In this case Theorem 4 below shows that (10) is also a solution for $z \in B$ and that this solution is *unique*. But for infinite state space I (10) is in general not the unique solution in B. The following theorem however indicates the form of the solution of (7).

Theorem 3

If $z \in B_+$, then any solution $Z \in B_+$ of (7) has the representation

$$Z = z + M * z + h, \tag{12}$$

where h satisfies

$$h = Q * h, \quad h \in B_+. \tag{13}$$

Proof. Suppose $Z \in B_+$ is a solution of (7). Then it follows by induction on n from (7) that

$$Z = \sum_{k=0}^{n} Q^{*k} * z + Q^{*(n+1)} * Z \tag{14}$$

where $Q^{*0} * z = z$. The sum in (14) is bounded by $Z \in B_+$ and increasing with n. Its limit exists and equals $M * z + z$ by Theorem 1 and this limit is also bounded by $Z \in B_+$ and therefore itself belongs to B_+. This implies also $M * z \in B$. Thus by Theorem 2 it follows that $M * z + z$ is a solution of (7).

The second term on the right hand side of (14) is decreasing with n and has a limit $h \in B_+$. This proves (12). (13) follows from $Z - h = z + M * z$ and the fact that $z + M * z$ and Z are both solutions of (7). ∎

This theorem shows that for $z \in B_+$ (10) may be considered as the *minimal* solution to (7). The following theorem now gives a sufficient condition that the solution (10) to the Markov renewal equation is *unique* and guarantees at the same time the *existence* of the solution not only for $z \in B_+$ but more generally for $z \in B$.

Theorem 4

If

$$\sup_i \sum_j Q_{ij}(t) < 1 \quad \text{for some } t > 0 \tag{15}$$

then $M * z \in B$ and

$$Z = z + M * z \in B \tag{16}$$

is the unique solution to (7) for any $z \in B$.

Proof. Let

$$F(t) = \sup_i F_i(t) = \sup_i \sum_j Q_{ij}(t).$$

For any i, j let Q_{ij}^{*n} be the i, j-element of Q^{*n}. Then for any i

$$\sum_j Q_{ij}^{*2} = \sum_j \left(\sum_k Q_{ik} * Q_{kj} \right)$$

$$= \sum_k Q_{ik} * \left(\sum_j Q_{kj} \right) = \sum_k Q_{ik} * F_k$$

$$\leqslant \sum_k Q_{ik} * F \leqslant F^{*2},$$

and by induction on n

$$\sum_j Q_{ij}^{*n} \leqslant F^{*n}. \tag{17}$$

Hence it follows from (8) that

$$\sum_j M_{ij}(t) = \sum_j \sum_{n=1}^{\infty} Q_{ij}^{*n}(t)$$

$$= \sum_{n=1}^{\infty} \sum_j Q_{ij}^{*n}(t) \leqslant \sum_{n=1}^{\infty} F^{*n}(t) = b(t).$$

But the last series is shown to be finite for any $t > 0$ by a similar argument as was used in the proof of Theorem 2.1.1, using the assumption that $F(t) < 1$ for some $t > 0$. So we have for all $i \in I$

$$\sum_j M_{ij}(t) \leqslant b(t) < \infty \quad \text{for any } t > 0. \tag{18}$$

For $z \in B$ let $z_i^+(t) = \max (z_i(t), 0)$ and $z_i^-(t) = \max (-z_i(t), 0), z^+ = (z_i^+),$ $z^- = (z_i^-), i \in I$. Then $z = z^+ - z^-$ and $z^+, z^- \in B_+$. (18) implies $M * z^+ \in B$ and $M * z^- \in B$ and hence by Theorem 2

$$z^+ + M * z^+ = z^+ + Q * (z^+ + M * z^+),$$

$$z^- + M * z^- = z^- + Q * (z^- + M * z^-).$$

subtracting the two equations shows that in fact $z + M * z$ is a solution of (7) and it belongs to class B.

To prove the uniqueness of this solution we have to show that (13) has only the trivial solution $h = 0$. From $h = Q * h, h \in B$ it follows that $h = Q^{*n} * h$ for $n = 1, 2, \ldots$. But for any $t \geqslant 0$

$$|h_i(t)| \leqslant \sum_k \int_0^t |h_k(t-x)| \, dQ_{ik}^{*n}(x)$$

$$\leqslant \sup_k \sup_{x \leqslant t} |h_k(x)| \sum_k Q_{ik}^{*n}(t),$$

where the last term is smaller than $F^{*n}(t)$ by (17). This term vanishes as $n \to \infty$ as a consequence of (18) and therefore $h = 0$ is the only solution to (13). ∎

Note that (15) is surely valid for *finite state spaces I* and Theorem 4 applies in this case. In order to clarify the meaning of (15), we prove the following theorem.

Theorem 5
(15) implies $L = \sup_n T_n = \infty$ (see (7.1.4)) with probability 1.

Proof. Let $h_{i,n}(t) = P[T_n \leqslant t | X_0 = i]$. Then by the proof of Theorem 1 and the proof of Theorem 4

$$h_{i,n}(t) = \sum_j Q_{ij}^{*n}(t) \to 0 \quad \text{as } n \to \infty.$$

But $h_i(t) = \lim_{n \to \infty} h_{i,n}(t) = P[\sup_n T_n \leqslant t | X_0 = i] = P[L \leqslant t | X_0 = i]$ and hence $L = \infty$ with probability 1. ∎

Examples
(e) *Transition probabilities of semi-Markov processes.* Consider Example (a) and suppose condition (15) is valid for the underlying Markov renewal process. Then in the Markov renewal equations of Example (a) we have $z_i(t) = 0$ for $i \neq j$ and $z_j(t) = (1 - F_j(t))$. Hence by (16)

$$p_{jj}(t) = (1 - F_j(t)) + \int_0^t (1 - F_j(t - x)) \mathrm{d}M_{jj}(x),$$

$$p_{ij}(t) = \int_0^t (1 - F_j(t - x)) \mathrm{d}M_{ij}(x)$$

is the unique solution of the Markov renewal equations. Formally $\mathrm{d}M_{ij}(x)$ may be interpreted as the probability of a renewal in state j at time x. Then the above formulae have an evident probabilistic interpretation. □

(f) *Continuous Markov chains.* In this case condition (15) is equivalent to the condition that the $q_i, i \in I$, are *bounded* (compare with Theorem 6.3.1). According to Example (e) we get

$$p_{ij}(t) = \delta_{ij} e^{-q_j t} + \int_0^t e^{-q_j(t - x)} \mathrm{d}M_{ij}(x)$$

as the solution to Kolmogorov's backward equations (see Example (b)). Differentiation with respect to t gives

$$\dot{p}_{ij}(t) = -q_j \delta_{ij} e^{-q_j t} - q_j \int_0^t e^{-q_j(t-x)} dM_{ij}(x) + \frac{dM_{ij}(t)}{dt}$$

$$= -q_j p_{ij}(t) + \dot{M}_{ij}(t).$$

$\dot{M}_{ij}(t) dt + o(dt)$ is the probability of a renewal in state j (starting from state i) during the interval from t to $t + dt$. But this event occurs if at time t the process is in some state $k \neq j$ and a jump from k to j occurs during the time interval dt. Hence it follows that (see (3.3.9))

$$\dot{M}_{ij}(t) = \sum_{k \neq j} p_{ik}(t) q_{kj}.$$

Introducing this into the formula for p_{ij} gives Kolmogorov's forward equations. □

(g) *M/G/1- and GI/M/1-queueing systems.* In the case of an *M/G/1*-queueing system the sojourn time distribution in state $i > 0$ is the service-time distribution $G(t)$ and for $i = 0$ the convolution of $G(t)$ with the exponential inter-arrival-time distribution (see Example (7.1.g)). So $F_i(t)$ has only two different forms, one for $i = 0$ and one for $i > 0$. Therefore condition (15) holds and the Markov renewal equations corresponding to *M/G/1*-Markov renewal processes have, according to Theorem 4, unique solutions. The same holds for the *GI/M/1*-system, as there $F_i(t)$ is equal for all i to the distribution $G(t)$ of inter-arrival times. □

By applying the key theorems of renewal theory (Theorems 2.3.3 and 2.3.4), the foregoing results allow the derivation of limits of solutions to Markov renewal equations. This is at least the case for *finite* state spaces I. In the case of infinite state spaces some supplementary technical problems arise. So the next theorem is limited to finite state spaces.

Theorem 6
Let $\{X_n, T_n, n = 0, 1, 2, \ldots\}$ be an irreducible aperiodic Markov renewal process with *finite* state space I. If $z \in B$ and each $z_i(t)$ is *directly Riemann integrable*, then for the unique solution $Z(t)$ of the Markov renewal equations (7)

$$\lim_{t \to \infty} Z_i(t) = \left\{ 1 \Big/ \sum_{j \in I} u_j \mu_j \right\} \sum_{i \in I} u_i \int_0^\infty z_i(t) dt, \tag{19}$$

where $u_i > 0, i \in I$, is a solution of

$$u_j = \sum_{i \in I} u_i p_{ij}, \quad j \in I, \tag{20}$$

and μ_j is the mean sojourn time in state j.

Proof. By Theorem 4, under the assumptions of the theorem the solution to (7) exists, is unique and is given by (16).

Then

$$\lim_{t\to\infty} Z_i(t) = \lim_{t\to\infty}\left\{ z_i(t) + \sum_{j\in I} \int_0^t z_j(t-x)\,\mathrm{d}M_{ij}(x) \right\}$$

$$= \lim_{t\to\infty}\left\{ z_i(t) + \int_0^t z_i(t-x)\,\mathrm{d}M_{ii}(x) \right\}$$

$$+ \sum_{\substack{j\in I \\ j\ne i}} \lim_{t\to\infty} \int_0^t z_j(t-x)\,\mathrm{d}M_{ij}(x). \qquad (21)$$

Summation and limit can be interchanged, because it is a finite sum. The limit of the first term equals

$$\frac{1}{\mu_{ii}} \int_0^\infty z_i(t)\,\mathrm{d}t \qquad (22)$$

by Theorem 2.3.3. $M_{ij}(t)$ is the renewal function of a delayed renewal process and it can be written as $M_{ij} = G_{ij} + G_{ij} * M_{jj}$ (by an obvious renewal argument). The last integral in (21) is therefore equal to $z_j * (G_{ij} + G_{ij} * M_{jj}) = (z_j * G_{ij}) + (z_j * (G_{ij} * M_{jj}))$ and can be written as

$$\varphi_j(t) + \int_0^t \varphi_j(t-x)\,\mathrm{d}M_{jj}(x)$$

with

$$\varphi_j(t) = \int_0^t z_j(t-x)\,\mathrm{d}G_{ij}(x).$$

If now $\varphi_j(t)$ is directly Riemann integrable, then it follows again from Theorem 2.3.3 that the limit of (22) as $t\to\infty$ equals

$$\frac{1}{\mu_{jj}} \int_0^\infty \varphi_j(t)\,\mathrm{d}t = \frac{1}{\mu_{jj}} \int_0^\infty \mathrm{d}t \int_0^t z_j(t-x)\,\mathrm{d}G_{ij}(x)$$

$$= \frac{1}{\mu_{jj}} \int_0^\infty \mathrm{d}G_{ij}(x) \int_0^\infty z_j(t)\,\mathrm{d}t = \frac{1}{\mu_{jj}} \int_0^\infty z_j(t)\,\mathrm{d}t,$$

because $G_{ij}(\infty) = 1$ by the irreducibility, and hence the recurrence of the finite state Markov renewal process (see Theorem 3.2.7). This together with (7.2.11) proves (19).

It remains to show that $\varphi_j(t)$ is in fact directly Riemann integrable. To show this, suppose first that $z_j(t)$ is a directly Riemann integrable *non-negative step function*. Fix $h > 0$, let $A_n = [nh - h, nh)$ and

$$z_j(t) = \sum_{n=1}^\infty a_n I_n(t) \geq 0 \quad \text{ for all } t \geq 0,$$

where $I_n(t)$ is the indicator function of the interval A_n ($I_n(t) = 1$ if $t \in A_n$, $I_n(t) = 0$ otherwise). Then

$$z_j * G_{ij}(t) = \sum_{k=1}^{n} a_k \{G_{ij}(t - kh + h) - G_{ij}(t - kh)\}.$$

Define furthermore

$$b_n = \sup_{t \in A_n} (z_j * G_{ij}(t)).$$

Now as

$$\sup_{t \in A_n} \{G_{ij}(t - kh + h) - G_{ij}(t - kh)\} \leqslant G_{ij}((n - k)h + h) - G_{ij}((n - k)h - h),$$

it follows that

$$b_n \leqslant \sum_{k=0}^{n-1} a_{n-k} \int_{kh-h}^{kh+h} dG_{ij}(x)$$

and thus

$$\sum_{n=1}^{\infty} b_n \leqslant \sum_{k=0}^{\infty} \int_{kh-h}^{kh+h} dG_{ij}(x) \sum_{n=k+1}^{\infty} a_{n-k}$$

$$\leqslant 2 \sum_{n=1}^{\infty} a_n < \infty$$

because $z_j(t)$ is directly Riemann integrable. This implies that $z_j * G_{ij}(t)$ is also directly Riemann integrable.

Now let $z_j(t)$ be an arbitrary directly Riemann integrable function, fix $h > 0$, define $\underline{m}_k$ and $\bar{m}_k$ according to (2.3.16) relative to $z_j(x)$ and $\underline{z}_j(t), \bar{z}_j(t)$ as the step functions with $a_k = \underline{m}_k$ or $a_k = \bar{m}_k$ respectively. Then

$$\underline{z}_j * G_{ij}(t) \leqslant z_j * G_{ij}(t) \leqslant \bar{z}_j * G_{ij}(t).$$

Since $\bar{z}_j(t) - \underline{z}_j(t)$ is a non-negative step function, it follows as above that

$$0 \leqslant h \sum_{n=1}^{\infty} b_n \leqslant 2h \sum_{n=1}^{\infty} (\bar{m}_n - \underline{m}_n)$$

with $b_n = \sup_{t \in A_n} ((\bar{z}_j - \underline{z}_j) * G_{ij}(t))$. But the series on the right hand side converges to 0 as $h \to 0$, because $z_j(t)$ is directly Riemann integrable. This shows that $\varphi_j(t) = z_j * G_{ij}(t)$ is in fact directly Riemann integrable. ∎

Note that Theorem 2.3.4 gives a sufficient condition for the $z_j(t)$ to be directly Riemann integrable. For infinite state spaces the situation is more complicated, as the interchange of limit and sum at the beginning of the last proof is not possible without further limitation of the class of functions $z(t)$. Let us note, however, without proof that Theorem 6 is valid for infinite state space I if $z \in B_+$, $z_j(t)$ is monotone decreasing for each $j \in I$ and

$$\sum_{j \in I} u_j z_j(0) < \infty, \qquad \int_0^t \sum_{j \in I} u_j z_j(t) \mathrm{d}t < \infty. \tag{23}$$

Examples

(h) *Limits of the transition probabilities of semi-Markov processes.*
Theorem 6 or its extension above applied to the Markov renewal equations of
Example (a) yields the same result as Theorem 7.2.4. So this is an alternative
way to derive the limits of the transition probabilities of a semi-Markov process. Note that also the result for the transient case (Theorem 7.2.2) can be
obtained from (16) using (2.3.1).   □

(i) *Limits of transition probabilities of processes with embedded Markov
renewal processes.* Let $\{Z_t, t \geqslant 0\}$ be a process as defined in Example (d). If
$r_{ij}(t) \geqslant 0$ is directly Riemann integrable, by Theorem 6 or its extension
above, it follows that

$$\lim_{t \to \infty} q_{ij}(t) = \lim_{t \to \infty} P[Z_t = j | X_0 = i]$$

$$= \frac{1}{\sum_{k \in I} u_k \mu_k} \sum_{k \in I} u_k \int_0^\infty r_{kj}(t) \mathrm{d}t.$$

These limits (if they exist) are independent of the initial state i. This result is
an extension of Theorem 7.2.5. It is somehow stronger than Theorem 7.2.5,
because it allows in some cases the explicit calculation of the limits more
easily than (7.2.18). See the next example.   □

(j) *M/G/1-queueing system.* The above example applies to the *M/G/1*-
queueing system. First Example (g) shows that the Markov renewal equations
of Example (d) have a unique solution in this special case. If $i > 0$, then T_1
equals the first service time and $P[T_1 > t] = 1 - G(t)$ and there is a transition
from i to $j \geqslant 0$ till time t if and only if there are exactly $j - i$ arrivals during
the interval $[0, t]$. Therefore

$$r_{ij}(t) = (1 - G(t))\frac{(\lambda t)^{j-i}}{(j-i)!} e^{-\lambda t} \quad \text{for } 0 < i \leqslant j.$$

$r_{00}(t)$ is simply the probability of no arrival in $[0, t]$. If finally $i = 0$ and $j > 0$
for $Z_t = j$ and $T_1 > t$, there must be an arrival at some time $s < t$, the service
starting at time s must last longer than $t - s$ and in that interval there must be
$j - 1$ further arrivals. Thus

$$r_{0j}(t) = \begin{cases} e^{-\lambda t} & \text{for } j = 0. \\ \int_0^t \lambda e^{-\lambda s}(1 - G(t-s))\dfrac{(\lambda(t-s))^{j-1}}{(j-1)!} e^{-\lambda(t-s)} \mathrm{d}s & j > 0. \end{cases}$$

$r_{00}(t)$ is clearly directly Riemann integrable. And $r_{ij}(t), i, j > 0$, is dominated by the monotone decreasing function $1 - G(t)$ and is therefore directly Riemann integrable, at least if the mean value of $G(t)$ is finite (which it must be, if the embedded Markov chain is recurrent; see Section 4.2). Finally $r_{0j}(t)$ for $j > 0$ is dominated by the convolution between the exponential inter-arrival distribution and $(1 - G(t))$. The end of the proof of Theorem 6 shows then that this convolution, and hence $r_{0j}(t)$, is directly Riemann integrable.

Now, if the mean number of arrivals r during a service time is less than 1, then a unique stationary distribution u_j of the embedded Markov chain exists (see Section 4.2). By the extension of Theorem 6 the limiting probabilities $q_j = \lim_{t \to \infty} q_{ij}(t)$ exist (see also Example (7.2.e)) and are given by

$$q_i = \left(1 \Big/ \sum_j u_j \mu_j\right) \sum_{i=0}^{j} u_i \int_0^\infty r_{ij}(t)\,dt,$$

at least if (23) is satisfied. Note that $\mu_j = S$ for $j \neq 0$, if S is the mean service time and $\mu_0 = S + 1/\lambda$, if λ is the arrival rate (see Section 4.2). Furthermore $r = \lambda S$ and therefore

$$\sum_j u_j \mu_j = 1/\lambda.$$

Let us compute the *generating function*

$$Q(z) = \sum_{j=0}^{\infty} q_j z^j$$

of the limiting probabilities q_j. Using the above formulae for q_j it follows that

$$Q(z) = \lambda u_0 \left\{ \int_0^\infty e^{-\lambda t}\,dt + \sum_{j=1}^{\infty} z^j \int_0^\infty dt \int_0^t \lambda e^{-\lambda s} \frac{(\lambda(t-s))^{j-1}}{(j-1)!} \right.$$

$$\left. \times e^{-\lambda(t-s)}(1 - G(t-s))\,ds \right\}$$

$$+ \lambda \sum_{i=1}^{\infty} u_i \sum_{j=i}^{\infty} z^j \int_0^\infty (1 - G(t)) \times \frac{(\lambda t)^{j-i}}{(j-i)!} e^{-\lambda t}\,dt$$

$$= \lambda u_0 \left\{ \int_0^\infty e^{-\lambda t}\,dt + z \int_0^\infty dt \int_0^t \lambda e^{-\lambda(t-s)}(1 - G(s))e^{-(\lambda - \lambda z)s}\,ds \right\}$$

$$+ \lambda \sum_{i=1}^{\infty} u_i z^i \int_0^\infty (1 - G(t))e^{-(\lambda - \lambda z)t}\,dt.$$

Using partial integration (see (1.3.36)) we obtain

$$Q(z) = u_0 \left\{ 1 + \frac{z}{1-z}(1 - g(\lambda - \lambda z)) \right\} + \sum_{i=1}^{\infty} \frac{u_i z^i}{1-z}(1 - g(\lambda - \lambda z))$$

$$= \frac{P(z)}{1-z}(1 - g(\lambda - \lambda z)) + u_0 g(\lambda - \lambda z),$$

where $g(z)$ is the *Laplace transform* of $G(t)$ (see 4.2.16)) and $P(z)$ the *generating function* of the stationary distribution u_j of the embedded Markov chain (see (4.2.13)). For the latter we have the Pollaczek–Khinchin formula (4.2.18) which we may introduce above to obtain

$$Q(z) = -\frac{g(\lambda - \lambda z)}{z - g(\lambda - \lambda z)} u_0(1 - g(\lambda - \lambda z)) + u_0 g(\lambda - \lambda z)$$

$$= \frac{(z - 1)g(\lambda - \lambda z)}{z - g(\lambda - \lambda z)} u_0 = P(z).$$

This shows that the limiting probabilities of the number of customers in the system are the *same* as those of the embedded Markov chain already discussed in Section 4.2. This confirms the remark made at the end of Example (4.2.*d*).

Notes to Chapter 7

Basic references to this chapter are the papers [1], [4] and [5]. Supplementary material on Markov renewal processes and semi-Markov processes may be found in the textbooks [2], [6].

The application of semi-Markov models in reliability theory is discussed in [3]. An important domain of application of Markov renewal methods is queueing theory; see for example [7]. Finally we refer to the optimization of Markov renewal processes; see [6], and compare also the notes to Chapter 5.

Bibliography

Chapter 1

[1] Chung, K. L. *A Course in Probability Theory*. New York, 1968.
[2] Doob, J. L. *Stochastic Processes*. New York, 1953.
[3] Feller, W. *An Introduction to Probability Theory and Its Applications*. New York, 1967 (vol. 1) and 1971 (vol. 2).
[4] Loève, M. *Probability Theory*. Princeton, N.J., 1963.
[5] Neveu, J. *Mathematical Foundations of the Calculus of Probability*. San Francisco, 1965.
[6] Parzen, E. *Modern Probability Theory and Its Applications*. New York, 1960.
[7] Rény, A. *Foundations of Probability*. San Francisco, 1970.

Chapter 2

[1] Barlow, R. E. & Proschan, F. *Mathematical Theory of Reliability*. New York, 1965.
[2] Cox, D. R. *Renewal Theory*. London and New York, 1962.
[3] Flehinger, B. J. A General Model for the Reliability Analysis Under Various Preventive Maintenance Policies. *Ann. Math. Stat.* **33** (1962), 137–56.
[4] Jardine, A. K. S. *Maintenance, Replacement and Reliability*. London, 1973.
[5] Jørgenson, D. W., McCall, J. J. & Radnor, R. *Optimal Replacement Policy*. Amsterdam, 1967.
[6] Kingman, J. F. C. *Regenerative Phenomena*. New York, 1972.
[7] Kozniewska. *Die Erneuerungstheorie; ökonomische Ersatzmodelle und ihre mathematische Lösung*. Berlin, 1969.
[8] Smith, W. L. Renewal Theory and Its Ramifications. *J. R. Stat. Soc.* **20** (1954) Ser. A, 9–48.
[9] Wolff, M. *Optimale Instandhaltungspolitiken in einfachen Systemen*. Berlin, Heidelberg, New York, 1970.
[10] Zelen, M. (ed.) *Statistical Theory of Reliability*. Wisconsin, 1963.

Chapter 3

[1] Chung, K. L. *Markov Chains with Stationary Transition Probabilities*. Berlin, Heidelberg, New York, 1967.

[2] Courtois, P. J. *Decomposability. Queueing and Computer System Applications*. New York, 1977.

[3] Dynkin, E. B. & Yushkevich, A. A. *Markov Processes: Theorems and Problems*. New York, 1969.

[4] Grassmann, W. K. Transient Solutions in Markovian Queueing Systems. *Comput. Op. Res.* **4** (1977), 47–53.

[5] Grassmann, W. K. Transient Solutions in Markovian Queues. *Eur. J. Op. Res.* **1** (1977), 396–402.

[6] Isaacson, D. L. & Madsen, R. W. *Markov Chains. Theory and Applications*. New York, 1976.

[7] Kemeny, J. G. & Snell, J. L. *Finite Markov Chains*. Princeton, 1960.

[8] Kemeny, J. G., Snell, J. L. & Knapp, A. W. *Denumerable Markov Chains*. Princeton, 1966.

Chapter 4

[1] Bhat, U. N. Sixty Years of Queueing Theory. *Management Sci.* **15** (1969), 280–94.

[2] Borovkov, A. A. *Stochastic Processes in Queueing Theory*. New York, Heidelberg, Berlin, 1976.

[3] Clarke, A. B. (ed.) *Mathematical Methods in Queueing Theory*. Berlin, Heidelberg, New York, 1974.

[4] Cohen, J. W. *The Single Server Queue*. Amsterdam, 1969.

[5] Cox, D. R. & Smith, W. C. *Queues*. London and New York, 1961.

[6] Ferschl, F. *Zufallsabhängige Wirtschaftsprozesse. Grundlagen und Anwendungen der Theorie der Wartesysteme*. Wien, Würzburg, 1964.

[7] Gnedenko, B. W. & Kowalenko, I. N. *Einführung in die Bedienungstheorie. München, Wien*, 1971.

[8] Gordon, W. J. & Newell, G. F. Closed Queueing Systems With Exponential Servers. *Op. Res.* **15** (1967), 254–65.

[9] Jackson, J. R. Networks of Waiting Lines. *Op. Res.* **5** (1957), 518–21.

[10] Jackson, J. R. Jobshop-Like Queueing Systems. *Management Sci.* **10** (1963), 131–42.

[11] Jaiswal, N. K. *Priority Queues*. New York, 1968.

[12] Kaufmann, A. & Cruon, R. *Les Phénomènes d'attente. Théorie et applications*. Paris, 1961.

[13] Kendall, D. G. Stochastic Processes Occurring in the Theory of Queues and Their Analysis by the Method of Imbedded Markov Chains. *Ann. Math. Stat.* **24** (1953), 338–54.

[14] Khinchin, A. Y. *Mathematical Methods in the Theory of Queueing*. London, 1960.

[15] Kingman, J. F. C. Inequalities in the Theory of Queues. *J. R. Stat. Soc.* **32** (1970) Ser. B, 102–10.

[16] Kleinrock, L. *Queueing Systems*, vols. 1, 2. New York, 1975.

[17] Lee, A. *Applied Queueing Theory*. St. Martin's Press, 1966.

[18] Le Gall, P. *Les Systèmes avec ou sans attente et les processus stochastiques*. Paris, 1962.

[19] Morse, P. *Queues, Inventories and Maintenance.* New York, 1960.
[20] Neuts, M. F. (ed.) *Algorithmic Methods in Probability.* Amsterdam, 1977.
[21] Reiser, M. Numerical Methods in Separable Queueing Networks. In: Neuts, M. F. *Algorithmic Methods in Probability.* Amsterdam, 1977.
[22] Schassberger, R. *Warteschlangen.* Wien, New York, 1973.
[23] Syski, E. *Congestion Theory.* New York, 1960.
[24] Takàcs, L. *Introduction to the Theory of Queues.* New York, 1960.

Chapter 5

[1] Beckmann, M. J. *Dynamic Programming of Economic Decisions.* Berlin, Heidelberg, New York, 1968.
[2] Bellman, R. E. *Dynamic Programming.* Princeton, 1957.
[3] Bühlmann, H., Loeffel, H. & Nievergelt, E. *Entscheidungs- und Spieltheorie.* Berlin, Heidelberg, New York, 1975.
[4] Derman, C. *Finite State Markovian Decision Processes.* New York, 1970.
[5] Fleming, W. H. & Rishel, R. *Deterministic and Stochastic Optimal Control.* New York, Heidelberg, Berlin, 1975.
[6] Hinderer, K. *Foundations of Non-Stationary Dynamic Programming With Discrete Time Parameter.* Berlin, Heidelberg, New York, 1970.
[7] Howard, R. A. *Dynamic Programming and Markov Processes.* Cambridge, Mass., 1960.
[8] Jewell, W. S. Markov-Renewal Programming. *Op. Res.* **11** (1963), 938–70.
[9] Kushner, H. J. *Introduction to Stochastic Control Theory.* New York, 1971.
[10] Lindley, D. *Making Decisions.* London, 1971.
[11] Mine, H. & Osaki, S. *Markovian Decision Processes.* New York, 1970.
[12] Nemhauser, G. L. *Introduction to Dynamic Programming.* London, New York, 1966.
[13] Neumann, K. *Dynamische Optimierung.* Mannheim, Wien, Zürich, 1969.
[14] Schneeweiss, Ch. *Dynamisches Programmieren.* Würzburg, Wien, 1974.
[15] White, D. J. *Dynamic Programming.* Edinburgh, London, 1969.

Chapter 6

[1] Bauknecht, K., Kohlas, J. & Zehnder, C. A. *Simulationstechnik.* Berlin, Heidelberg, New York, 1976.
[2] Crane, M. A. & Lemoine, A. J. *An Introduction to the Regenerative Method for Simulation Analysis.* Berlin, Heidelberg, New York, 1977.
[3] Ermakow, S. M. *Die Monte-Carlo-Methode und verwandte Fragen.* München, Wien, 1975.

[4] Fishman, G. S. *Concepts and Methods in Discrete Event Digital Simulation.* New York, 1973.

[5] Freiberger, W. & Grenander, U. *A Short Course in Computational Probability.* New York, Heidelberg, Berlin, 1971.

[6] Hammersley, J. M. & Handscomb, D. C. *Monte-Carlo-Methods.* London, 1964.

[7] Jansson, B. *Random Number Generators.* Stockholm, 1966.

[8] Kleijnen, J. P. C. *Statistical Techniques in Simulation*, vols. 1, 2. New York, 1974.

[9] Knuth, D. E. *The Art of Computer Programming: Seminumerical Algorithms*, vol. 2. Reading, Mass., 1969.

[10] Lewis, T. G. *Distribution Sampling for Computer Simulation.* Lexington, Mass., 1975.

[11] Schmitz, N. & Lehmann, F. *Monte-Carlo-Methoden I. Erzeugen und Testen von Zufallszahlen.* Meisenheim am Glan, 1976.

Chapter 7

[1] Çinlar, E. Markov Renewal Theory. *Adv. appl. Prob.* **1** (1969), 123–87.

[2] Çinlar, E. *Introduction to Stochastic Processes.* Englewood Cliffs, N. J., 1975.

[3] Gaede, K. W. *Zuverlässigkeit. Mathematische Modelle.* München, Wien, 1977.

[4] Pyke, R. Markov Renewal Processes. Definitions and Preliminary Properties. *Ann. Math. Stat.* **32** (1961), 1231–42.

[5] Pyke, R. Markov Renewal Processes with Finitely Many States. *Ann. Math. Stat.* **32** (1961), 1243–59.

[6] Ross, S. M. Applied Probability Models with Optimization Applications. San Francisco, 1970.

[7] Teghem, J., Loris-Teghem, J. & Lambotte, J. P. *Modèles d'Attente M/G/1 et GL/M/1 à Arrivées et Services en Groupe.* Berlin, Heidelberg, New York, 1969.

Index